Rational Design of Stable Protein Formulations

Theory and Practice

Pharmaceutical Biotechnology

Series Editor: Ronald T. Borchardt
The University of Kansas
Lawrence, Kansas

Recent volumes in this series:

Volume 7 PHYSICAL METHODS TO CHARACTERIZE PHARMACEUTICAL PROTEINS
Edited by James N. Herron, Win Jiskoot, and Daan J. A. Crommelin

Volume 8 MODELS FOR ASSESSING DRUG ABSORPTION AND METABOLISM
Edited by Ronald T. Borchardt, Philip L. Smith, and Glynn Wilson

Volume 9 FORMULATION, CHARACTERIZATION, AND STABILITY OF PROTEIN DRUGS: Case Histories
Edited by Rodney Pearlman and Y. John Wang

Volume 10 PROTEIN DELIVERY: Physical Systems
Edited by Lynda M. Sanders and R. Wayne Hendren

Volume 11 INTEGRATION OF PHARMACEUTICAL DISCOVERY AND DEVELOPMENT: Case Histories
Edited by Ronald T. Borchardt, Roger M. Freidinger, Tomi K. Sawyer, and Philip L. Smith

Volume 12 MEMBRANE TRANSPORTERS AS DRUG TARGETS
Edited by Gordon L. Amidon and Wolfgang Sadée

Volume 13 RATIONAL DESIGN OF STABLE PROTEIN FORMULATIONS: Theory and Practice
Edited by John F. Carpenter and Mark C. Manning

Volume 14 DEVELOPMENT AND MANUFACTURE OF PROTEIN PHARMACEUTICALS
Edited by Steven L. Nail and Michael J. Akers

A Chronological Listing of Volumes in this series appears at the back of this volume

---

A Continuation Order Plan is available for this series. A continuation order will bring delivery of each new volume immediately upon publication. Volumes are billed only upon actual shipment. For further information please contact the publisher.

Rational Design of Stable Protein Formulations

Theory and Practice

Edited by

John F. Carpenter

and

Mark C. Manning

University of Colorado Health Sciences Center
Denver, Colorado

Kluwer Academic / Plenum Publishers
New York, Boston, Dordrecht, London, Moscow

Library of Congress Cataloging-in-Publication Data

Rational design of stable protein formulations: theory and practice/edited by John F. Carpenter, Mark C. Manning.
p. cm. — (Pharmaceutical biotechnology; v. 13)
Includes bibliographical references and index.
ISBN 0-306-46741-0
1. Protein drugs—Stability. 2. Protein engineering. 3. Drugs—Design. I. Carpenter, John F. II. Manning, Mark C. III. Series.
RS431.P75 R38 2002
615′.19—dc21

2001057997

ISBN: 0-306-46741-0

233 Spring Street, New York, N.Y. 10013

http://www.wkap.nl/

10 9 8 7 6 5 4 3 2 1

A C.I.P. record for this book is available from the Library of Congress

Printed in the United States of America

Contributors

Tsutomu Arakawa • Alliance Protein Laboratories, Thousand Oaks, California 91360

John Carpenter • Center for Pharmaceutical Biotechnology, Department of Pharmaceutical Sciences, School of Pharmacy, University of Colorado Health Sciences Center, Denver, Colorado 80262

Byeong S. Chang • Department of Pharmaceutics and Drug Delivery, Amgen, Inc., Thousand Oaks, California 90132

William Garzon-Rodriguez • Center for Pharmaceutical Biotechnology, Department of Pharmaceutical Sciences, School of Pharmacy, University of Colorado Health Sciences Center, Denver, Colorado 80262

Susan Hershenson • Department of Pharmaceutics and Drug Delivery, Amgen, Inc., Thousand Oaks, California 90132

Bert Ho • Tularik Corporation, S. San Francisco, California 94080

LaToya S. Jones • Center for Pharmaceutical Biotechnology, Department of Pharmaceutical Sciences, School of Pharmacy, University of Colorado Health Sciences Center, Denver, Colorado 80262

Brent S. Kendrick • Amgen, Inc., Boulder, Colorado 80301

Geoffrey Lee • Department of Pharmaceutical Technology, Friedrich-Alexander University, Erlangen, Germany

Tiansheng Li • Amgen, Inc., Thousand Oaks, California 90132

Mark Manning • Center for Pharmaceutical Biotechnology, Department of Pharmaceutical Sciences, School of Pharmacy, University of Colorado Health Sciences Center, Denver, Colorado 80262

Jeffrey D. Meyer • Center for Pharmaceutical Biotechnology, Department of Pharmaceutical Sciences, School of Pharmacy, University of Colorado Health Sciences Center, Denver, Colorado 80262

Linda O. Narhi • Amgen, Inc., Thousand Oaks, California 90132

Rajiv Nayar • HTD Biosystems, Hercules, California 94547

Theodore W. Randolph • Center for Pharmaceutical Biotechnology, Department of Chemical Engineering, University of Colorado, Boulder, Colorado 80503

Preface to the Series

A major challenge confronting pharmaceutical scientists in the future will be to design successful dosage forms for the next generation of drugs. Many of these drugs will be complex polymers of amino acids (e.g., peptides, proteins), nucleosides (e.g., antisense molecules), carbohydrates (e.g., polysaccharides), or complex lipids.

Through rational drug design, synthetic medicinal chemists are preparing very potent and very specific peptides and antisense drug candidates. These molecules are being developed with molecular characteristics that permit optimal interaction with the specific macromolecules (e.g., receptors, enzymes, RNA, DNA) that mediate their therapeutic effects. Rational drug design does not necessarily mean rational drug delivery, however, which strives to incorporate into a molecule the molecular properties necessary for optimal transfer between the point of administration and the pharmacological target site in the body.

Like rational drug design, molecular biology is having a significant impact on the pharmaceutical industry. For the first time, it is possible to produce large quantities of highly pure proteins, polysaccharides, and lipids for possible pharmaceutical applications. Like peptides and antisense molecules, the design of successful dosage forms for these complex biotechnology products represents a major challenge to pharmaceutical scientists.

Development of an acceptable drug dosage form is a complex process requiring strong interactions between scientists from many different divisions in a pharmaceutical company, including discovery, development, and manufacturing. The series editor, the editors of the individual volumes, and the publisher hope that this new series will be particularly helpful to scientists in the development areas of a pharmaceutical company, (e.g., drug metabolism, toxicology, pharmacokinetics and pharmacodynamics, drug delivery, preformulation, formulation, and physical and analytical chemistry). In addition, we hope this series will help to build bridges between the development scientists and scientists in discovery (e.g., medicinal chemistry, pharmacology, immunology, cell biology, molecular biology) and in manufacturing (e.g., process chemistry, engineering). The design of successful dosage forms for the next generation of drugs will require not only a high level of expertise by individual scientists, but also a high degree of interaction between scientists in these different divisions of a pharmaceutical company.

Finally, everyone involved with this series hopes that these volumes will also be useful to the educators who are training the next generation of pharmaceutical scientists. In addition to having a high level of expertise in their respective disciplines, these young scientists will need to have the scientific skills necessary to communicate with their peers in other scientific disciplines.

RONALD T. BORCHARDT
Series Editor

Preface

The need to improve drug development approaches has become even more important in this post-genomic era. The number of drug candidates is escalating in most companies. As a result, every aspect of drug development must become more efficient and accurate. This volume is intended as a treatise on the state-of-the-art in formulation and stabilization of protein pharmaceuticals. As we approach the twentieth anniversary of the introduction of a recombinant protein therapeutic agent (human insulin, 1982), it is timely to summarize the advances made in formulation of proteins. The earlier two volumes in this series dedicated to protein stability continue to be widely used in the industry as reference works. However, this volume is intended to move beyond that work. Rather than providing a traditional description of mechanisms and concepts, this volume is intended to be more practical, trying to emphasize that our understanding of protein stability has evolved to the point where rational design of some formulations can be achieved. Furthermore, two of the chapters introduce new stabilization and formulation strategies that may provide future directions in this field. We believe the modern practitioner of protein formulation should find this volume of use, whether they are experts or neophytes.

The first chapter, authored by Byeong Chang and Susan Hershenson from Amgen, is entitled "Practical Approaches to Protein Formulation Development". This chapter will not only talk about modes of decomposition of protein but provides readable and practical guide on how to turn theoretical insight into a manufacturable, stable product. It also provides the reader with insight into the practical constraints placed on scientists actually doing formulation development.

Chapter 2, by Tsutomu Arakawa, Tiansheng Li, and Linda Nahri, introduces the reader to critical issues of refolding proteins produced in *E. coli.* For someone new to this field, the authors provide a number of experimental details and clearly describe the current approaches towards obtaining high yields of native proteins from inclusion bodies. By providing such great detail, the reader can quickly identify the prior work that is most relevant to their current needs.

The third chapter, entitled "Physical Stabilization of Proteins in Aqueous Solution", is written by Brent Kendrick, Tiansheng Li, and Byeong Chang. A fundamental principle in protein stabilization is the ability of additives (excipients) to alter the thermodynamic stability of a protein's structure. These thermodynamic effects arise from preferential binding or exclusion of the additive from the surface of the protein, as first described by Serge Timasheff and co-workers. This chapter provides a detailed tutorial on the theory and then proceeds to outline the impact of these effects on the stability of protein pharmaceuticals. These effects are important in development of liquid formulations, stabilization during lyophilization, and protein aggregation.

The next chapter introduces an emerging appreciation that susceptibility to chemical degradation can be modulated by the conformation of the protein (or even peptide). Entitled "Effects of Conformation on the Chemical Stability of Pharmaceutically Relevant Peptides", by Jeffrey Meyer, Bert Ho, and Mark Manning, summarizes the data in the literature that indicate it is possible to minimize chemical degradation by altering conformation. One possible method for accomplishing this would be to add preferentially excluded solutes (as described in chapter 3) to compact the structure of the protein. This has worked to diminish aggregation and is now being shown to be effective at slowing chemical decomposition as well. Altogether, this approach could provide a new strategy for developing liquid formulations of peptides and proteins.

The one area where our understanding of protein behavior has truly advanced is for freeze-dried formulations. The authors of the next chapter, John Carpenter, Byeong Chang, William Garzon-Rodriguez, and Theodore Randolph, provide an overview of what is required to produce a stable formulation of a protein in a lyophilized cake. Not only do the authors provide an excellent summary of the advances that have been made in understanding the behavior of proteins during freeze-drying, they translate that theoretical insight into a step-by-step guide to producing reasonable, practical formulations that should work for most any protein pharmaceutical.

Chapter 6, authored by Geoff Lee, is an overview of the development of spray-dried protein formulations. This process has received a large amount of attention in recent years, and this chapter describes the process, the equipment, and formulation concerns that need to be addressed if one is to use this approach to make dried powders of proteins that will retain native structure and activity.

Chapter 7 focuses on the role of surfactants in stabilizing proteins. Over the past few years, an increasing number of mechanisms by which surfactants can stabilize proteins have been described. This chapter, by Theodore Randolph and LaToya Jones, provides an overview about the behavior of this important class of excipients. Even the most experienced formulation scientist probably does not appreciate all of the subtleties of protein-surfactant interactions. Understanding these mechanisms will allow for a more rational choice of which surfactant (if any) to use and at what concentration.

The last chapter is intended to spur discussion about new strategies to designing formulations of protein pharmaceuticals. Although the drug discovery field has been revolutionized by combinatorial libraries, high throughput screening, and genomics, similar high throughput approaches have not received much attention in the drug development field. This chapter, written by Rajiv Nayar and Mark Manning, will hopefully cause workers in this field to examine their current approaches and seek to find more efficient modes of operation. It is the authors' belief that the demand to develop more products simultaneously with less material and in less time will ultimately transform the field of formulation science.

For anyone who wishes to learn the state-of-the-art in formulating and stabilizing protein drug products, this book should provide information on all aspects, covering not only what is well established, but also where we are and where we are going. The goal to ensure that the reader has the mechanistic understanding of how proteins degrade and can be stabilized rationally, and then provide practical advice on how to translate that knowledge into action.

JOHN CARPENTER
MARK C. MANNING

Contents

Chapter 1

Practical Approaches to Protein Formulation Development
Byeong S. Chang and Susan Hershenson

Chapter 2

Recombinant Production of Native Proteins from *Escherichia coli*
Tsutomu Arakawa, Tiansheng Li, and Linda O. Narhi

Chapter 3

Physical Stabilization of Proteins in Aqueous Solution
Brent S. Kendrick, Tiansheng Li, and Byeong S. Chang

Chapter 4

Effects of Conformation on the Chemical Stability of Pharmaceutically Relevant Polypeptides
Jeffrey D. Meyer, Bert Ho, and Mark C. Manning

Chapter 5

Rational Design of Stable Lyophilized Protein Formulations: Theory and Practice
John F. Carpenter, Beyong S. Chang, William Garzon-Rodriguez, and Theodore W. Randolph

Chapter 6

Spray-Drying of Proteins
Geoffrey Lee

Chapter 7

Surfactant-Protein Interactions
Theodore W. Randolph and LaToya S. Jones

Chapter 8

High Throughput Formulation: Strategies for Rapid Development of Stable Protein Products
Rajiv Nayar and Mark C. Manning

1

Practical Approaches to Protein Formulation Development

Byeong S. Chang and Susan Hershenson

INTRODUCTION

As is the case with other pharmaceuticals, formulation development is one of the critical steps in developing a protein as a therapeutic product. Development of stable protein formulations may require even more resources and effort than conventional small molecule pharmaceuticals. Proteins typically have more stability issues as a result of their complexity and delicate structural stability. Fortunately, a great deal of research regarding protein stability has been conducted and this information is readily available in the literature (reviewed by Manning et al., 1989; Chen, 1992; Ahern and Manning, 1992a, 1992b; Arakawa et al., 1993; Cleland et al., 1993; Wang and Pearlman, 1993; Pearlman and Wang, 1996; Volkin and Middaugh, 1997). Ultimately, it would be ideal to be able to develop a pure pharmaceutical containing only the native protein. However, it is not practical to have only the native form of a protein in the formulation because the protein must be purified from a complex biological mixture containing a pool of other proteins which includes misfolded, denatured, and degraded forms of the same protein. Furthermore, a major challenge is to maintain the integrity of the purified protein during routine pharmaceutical processing, storage, handling, and delivery to the patient. One could envision achieving this goal by developing a formulation with

Byeong S. Chang and Susan Hershenson • Department of Pharmaceutics and Drug Delivery, Amgen, Inc., Thousand Oaks, CA 90132.

Rational Design of Stable Protein Formulations, edited by Carpenter and Manning. Kluwer Academic / Plenum Publishers, New York, 2002.

perfect stability, i.e., no physical and chemical change in the protein. Because proteins are complex molecules composed of numerous reactive chemical groups and delicate three-dimensional structures, identifying a set of conditions to keep all components stable is virtually impossible. In general, commercial therapeutic protein formulations are developed under the assumption that some degree of physicochemical changes will occur during storage and handling.

Realizing that it is impossible to develop a perfectly stable formulation, especially while meeting an aggressive product development timeline, the main objective then becomes one of maintaining the appropriate safety and efficacy of the product. In order to achieve this objective, it is imperative to understand the broad spectrum of degradation pathways affecting proteins, and to have available equipment and expertise in an extensive repertoire of analytical methods. Formulation development focuses on determining the potential degradation pathways, assessing the significance of each and optimizing variables to minimize the degradation products that are clinically significant.

Regulatory guidelines also are critical elements for guiding formulation development. They provide information about how to conduct studies and obtain useful results for evaluating formulations. The results obtained allow formulation scientists to write an appropriate developmental pharmaceutics section in regulatory filings. The guidelines also help to evaluate the significance of some inevitable degradation products that are produced during manufacturing, shipping and storage. For example, if the degradation products have properties comparable to those of the desired product with respect to activity, efficacy, and safety, they can be classified as product-related substances. This classification is significantly different from considering the degradation product as an impurity when there is not sufficient supporting evidence to justify classification as a product-related substance (Appendix, Regulatory Document 1). Understanding of these practical issues of regulatory requirements is critical for formulation scientists during design, implementation, evaluation and reporting of their studies.

In addition to insight into the scientific and regulatory issues, developing commercial formulations requires a clear understanding of the potential market. For example, indication, patients, method of delivery, frequency of dosing, typical dose requirement, market distribution and other business-related information will provide directions for the design of a successful formulation. Also, it is important to consider the competitiveness of the formulation as compared to other products available in the market.

In this chapter, an overview of critical factors affecting the design of therapeutic protein formulations and a general guide to developing commercially viable dosage forms for protein pharmaceuticals will be discussed. Since the majority of practical issues are not covered very well in the scientific literature, this chapter also includes information from regulatory guidance documents (see Appendix), labels from marketed products and routine industrial experience.

PREPARATION FOR FORMULATION DEVELOPMENT

Resource Requirements for Formulation Development

A list of resources that should be available before starting formulation development is summarized in Table 1. Some resources may not be as critical as others, depending on the nature of problems encountered, but it is important to have sufficient resources to discover major formulation issues as early as possible in the product development process. Without the appropriate equipment and personnel, the development of an acceptable formulation can be greatly hampered, potentially to such a degree that the product is never brought to market. Of course, not all of the resources need to be at the company developing the therapeutic protein, but if outside contractors are employed it is essential that timely access to resources is available. In addition, the quality of raw materials should be carefully evaluated because unexpected impurities may introduce unnecessary

Table 1.
Resource Requirements for Initial Protein Formulation Development

Resources	Requirement	Example
Purified protein	Representative of manufacturing process; sufficient quantity to cover dose bracket, formulation variables, and stress conditions; minimum complication by impurity (precipitation of impurity, degradation by impurity like proteolytic cleavage).	Purified bulk, sample from final purification process
Qualified excipients	Pharmaceutically acceptable quality, manufacturers with qualified production procedures and sufficient scale, specifications on critical impurities, quality that can be carried on to clinical studies and commercial distribution.	USP, Ph. Eur, JP
Access to fill finish facility	Capability to sterilize container/closure components; fill/finish under aseptic environment; head-space purge system; drying equipment	Sterile hood, filling machine, lyophilizer
Analytical instruments	Structural analyses; concentration determination, chromatographic analyses; electrophoresis; bioassays; other microcharacterizations.	CD, UV, fluorescence, HPLCs, mass-spectrometry, SDS-PAGE
Facility to accommodate stability studies	Controlled temperature, controlled light exposure, controlled relative humidity, devices to provide controlled agitation.	Freezer, refrigerator, incubator, light chamber, RH incubator, agitator

Table 2.
Examples of Information Useful for Designing Formulation Studies

Information	Examples
Clinical indication	Site of treatment (self-administration, office visit, hospital), methods of delivery, concomitant medication, competition.
Patient population	Age, strength, tolerability, capability to manipulate devices, sensitivity to excipient
Typical routes of delivery	Injectables (IV, SC, IM, IP, ICV, IT, IO), topical, inhalation, nasal, oral, etc.
Dose requirement	PK profile, frequency of dosing, variable vs. fixed dose, single-dose/multidose
Drug interaction	Co-administration with other drug, dilution or reconstitution with other solution; presence of undesirable compounds like reducing sugars, preservatives
Typical dosage forms	Liquid, lyophilized, spray-dried, aerosol by liquid or powder, other novel carrier; stability, physical properties, reconstitution art
Container/closure	Vial/stoppers, prefilled syringes, prefilled cartridges, dual-chamber cartridges, blister packages, product contact material, leacheates, breakage, light sensitivity, moisture penetration
Delivery device	Syringes, prefilled-syringes, pen injectors, auto-injectors, needle-free injectors, inhalation devices, infusion pumps

complications in the stability profile that is determined during formulation development and testing.

Useful Information for Designing Formulations

The configuration of a protein formulation is affected by how the drug will be used as a product. If such information is available when designing formulation studies, it is recommended to consider the limitations and challenges associated with each application. Examples of such information are listed in Table 2.

PREFORMULATION DEVELOPMENT

It is important to understand the critical properties of a protein before starting large studies to design and test the final formulation. Preformulation studies are designed to learn about the protein's susceptibility to a variety of pharmaceutically relevant stresses. The main objectives of preformulation research

Table 3.
Information Obtained from Pre-formulation Studies for Protein Pharmaceuticals

Characterization	Examples
Physical properties	Primary, secondary, tertiary and quaternary structures, solubility, viscosity, self-association, hydrophobicity, molecular weight, extinction coefficient, glycosylation, effects of ionic strength, etc.
Biological properties	Substrate or receptor affinity, in vitro activity model, *in vivo* preclinical model, etc.

include: general characterization of the product; investigation of potential stability issues; development of relevant analytical methods; establishment of a stability profile with stability-indicating assays; and identification of major formulation challenges. A summary of these issues will be presented in this section.

Characterization of Protein Pharmaceuticals

There is available in the literature extensive coverage of analytical methods and their principles for the characterization of proteins (e.g., Jones, 1993; Reubsaet et al., 1998; Herron et al., 1995). General points to consider when characterizing a protein as an active ingredient are listed in Table 3. As a protein's biological activity is dependent on its structure, significant emphasis has been given to structural properties and stability against various stresses.

Accelerated Stability Studies

In order to predict potential stability problems within a short period of time and to develop appropriate analytical methods, proteins are exposed to stronger-than-real stresses and various degradation products induced by the stresses are examined. The results obtained from these so-called "accelerated stability studies" might also be useful to predict the kinetics of the degradation processes under real handling conditions, when there are not sufficient real-time results available because of time and resource constraints. However, the accelerated stability study is not acceptable to determine expiry of the product, so it is best used to rank order the importance of different degradation pathways. Approaches and cautions in the extrapolation of data from accelerated stability testing to real-time stability and normal handling conditions are discussed in detail below.

Table 4.
Various Conditions Used to Accelerate Protein Degradation

Stresses	Routine ranges	Practical applications	Problems to monitor
Temperature	0–50°C	Storage, shipping, handling, delivery	Structural changes (precipitation, aggregation, recovery loss), solubility, increased reaction rates for all degradations
Light	>1.2 million lux hrs illumination, >200 watt hrs/square meter UV energy	Light exposure, container, package	Oxidation, cleavage
Freezing	Multiple freeze-thaw, liquid nitrogen freeze	Frozen storage, accidental freezing, lyophilization	Precipitation, aggregation, pH change, crystallization of excipients
Oxidation	Oxygen purge, peroxide spike	Storage, excipient stability, impurity	Oxidations, inactivation
Humidity	0–100% RH	Storage, container integrity, powder	Moisture content, moisture related degradations
Mechanical stresses	Vortex, agitation, shear-stress ($3000\,s^{-1}$)	Manufacturing, filling, shipping, handling, delivery	Precipitation, aggregation, recovery loss
Other denaturants	—	Impurities, pH, denaturing excipients	Precipitation, aggregation, recovery loss, structural changes

Conditions to accelerate various degradation reactions in protein products and potential problems to monitor are listed in Table 4. Proteins contain numerous amino acid side chains and delicate three-dimensional structures, which can be susceptible to different stresses. Therefore, it is important to test the protein under a variety of physical and chemical stresses in order to provide a good simulation of the degradation products that can be generated.

Development of Analytical Methods

A brief summary of typical analytical methods is presented in Table 5. It is essential to have a wide range of analytical methods available to identify and characterize degradation products. The analytical methods should be further selected and customized to accommodate the specific needs for each protein product.

Table 5.
Typical Methods Used to Characterize Proteins and Degradation Products

Methods	Examples	Applications
Column chromatography	HPLC, FPLC, low pressure LC; size-exclusion, reversed-phase, ion-exchange, hydrophobic, affinity columns; coupled with UV, fluorescence, RI, and other analytical instruments as detectors	Most physical and chemical degradations, excipient impurities, leacheates
Electrophoresis	SDS-PAGE, native PAGE, isoelectric focusing, capillary electrophoresis, etc.	Degradations with changes in size and/or charge
Spectroscopy	CD, fluorescence, FTIR, UV, Raman, NMR, etc.	Structural changes, chemical modifications of side groups
Thermal analysis	Differential scanning calorimetry, thermogravimetric analysis, thermomechanical analysis, etc.	Protein structure, lyophilized cake structure, powder characterizations
Light scattering/turbidity	Dynamic light scattering, other light scattering devices, turbidity, particle size determination, particle counter, etc.	Aggregation, precipitation, molecular weight determination
Other microcharacterization methods	Peptide mapping, peptide sequencing, amino acid analysis, mass spectrometry, other specific analyses for individual reactive groups	Identification of impurities and chemical degradation, analysis of complex proteins, e.g., antibody and glycoprotein

Evaluation of the Significance of Problems

As stated earlier, it would be ideal to have a pure protein in an absolutely stable formulation. In reality, however, scientists have to design formulation based on compromises that deal with several different potential problems. To make matters worse, the formulation needs to be recommended long before it is evaluated fully, because of typical aggressive timelines in the industry. In general, the formulation will have to be optimized based on assumptions and extrapolations of results obtained during a limited time. Therefore, it is important to utilize the given time efficiently, by collecting as much relevant information as possible for evaluating the significance of each problem.

Due to the marginal stability of proteins, it is possible to create rapidly a variety of degradation products during accelerated stability testing. However, not all the degradation products that are observed will be significant under normal handling, shipping and storage conditions. Furthermore, the rank order of different degradation processes under accelerated stability testing will NOT be the same under practical handling conditions, because each reaction has a different temperature dependency, i.e., reaction order and activation energy. Another important thing to keep in mind when evaluating different degradation processes is the contribution to the pharmaceutical quality of the product. Critical degradation products should be designated not on the quantity obtained during accelerated stability studies, but on a comprehensive understanding of their contributions to the quality of the product.

Quantitative Assessment. Usually, the rate of each degradation reaction during real storage conditions (e.g., at 4–8°C) should be very slow. In order to predict the rate under real storage conditions, reaction rates obtained under accelerated stability conditions can be extrapolated by using the Arrhenius equation.

Predicting the correct reaction rate requires a proper understanding of the reaction order, because the amount of the degradation product does not linearly increase over time unless the reaction follows zero-order kinetics. One way to determine the reaction order is to calculate the linearity between the concentration of residual native protein and time. The reaction order can be calculated from:

$$dC/dt = kC^n \tag{1}$$

where, C is the protein concentration, t is the time, k is the reaction rate constant, n is the reaction order. For a zero order reaction $n = 0$ and $C_0 - C = k \cdot t$. For a first order reaction $n = 1$ and $\log C_0 - \log C = k \cdot t$. For a second order reaction $n = 2$ and $1/C - 1/C_0 = k \cdot t$. For each reaction order a characteristic transformation of concentration will show a linear relationship with time, e.g., log C will show a linear relationship with time for a first order reaction, whereas $1/C_0$ will be linear versus time for second order reactions.

After obtaining the reaction rate constants for different temperatures, the activation energy can be found by using with the Arrhenius equation.

$$k = A \cdot e^{(-Ea/RT)} \tag{2}$$

The activation energy is obtained by plotting (log k) vs (1/T) and determining the slope of the plot. Using the activation energy, the reaction rate constant for real storage condition, e.g., 2–8°C for refrigerated storage, can be estimated by extrapolation.

Care must be taken when extrapolating the rate constants because there are numerous cases when the Arrhenius relationship does not apply. The Arrhenius reaction applies only to irreversible reactions where the product is accumulated as a single quantifiable species. If the degradation product is the result of serial reactions, the reaction order and rate constant will be determined only by the rate-limiting reaction at the testing condition. If the rate-limiting reaction changes or another significant variable is introduced at the extrapolated condition, then the extrapolated rate constant will be in error. Such complications routinely can be found in various degradation pathways that are affected by temperature-sensitive reactions such as structural changes, pH changes, by physical changes such as the glass transition of an amorphous phase in a lyophilized formulation and by changes in reactants like dissolved oxygen. For example, proteins have relatively high activation energies for structural changes. Thus, many reactions that are dependent on a major perturbation of the native protein structure tend to cause substantial damage under accelerated storage conditions, e.g., higher temperature. However, many of these reactions are not necessarily major problems when the product is stored at 2–8°C where the protein maintains its native conformation. In contrast, degradation reactions with lower activation energies, which might not be coupled to protein conformational changes (e.g., oxidation of surface methionine residues), tend to be much more problematic in the development of protein formulations.

Results from stresses (e.g., agitation) other than heat imposed in accelerated stability studies potentially could be extrapolated in a similar way to storage and actual handing conditions. The major challenge is the ability to assess quantitatively the magnitude of the stress so that the stress-stability relationship can be established. To date there are not published guidelines for such quantitation.

Qualitative Assessment. Not all of the degradation products are equivalent in terms of their contribution to the pharmaceutical quality of the protein product. Therefore, a qualitative assessment of each degradation product is important when weighing their significance. Regardless of their quantity, some degradation products are generally less acceptable than others. If the degradation product comprises the safety of the product, then it should be considered less acceptable. For example, even at levels of a few percent or less of the total protein population, non-native aggregates can cause adverse reactions in patients such as immune responses and even anaphylactic shock. Also, if the degradation reaction results in the inactivation of protein, it needs to be considered more important than other degradation reactions that do not affect the activity. For example, certain chemical changes (e.g., deamidation) may not alter the activity of a given product, whereas other reactions (e.g., oxidation) may render that product inactive. The relative impact of each degradation product on safety and efficacy cannot be predicted and must be determined for each protein therapeutic.

Information useful for the qualitative evaluation of degradation products includes the identity, clinical and preclinical experiences with the degradation product, biological activity, stability and potential side effects. Regulatory documents also provide clear guidance about what assessment is necessary for protein degradation products (Appendix, Regulatory Documents 2,3). The criteria for identification, reporting and qualification of degradation products are set up based on the total patient exposure and percentage of the degradation products (Appendix, Regulatory Document 3). When certain degradation products cannot be prevented from forming in the formulation, their inclusion in the product must be qualified, which may require that additional information on safety and efficacy be obtained as recommended in the regulatory guidelines (Appendix, Regulatory Documents 2,3). Obtaining this information requires a significant amount of additional work, which may include clinical studies.

FORMULATION DEVELOPMENT

As discussed above, the critical parameters affecting the pharmaceutical quality of a protein therapeutic are defined during the preformulation studies. In formulation development, the effects of formulation variables on the defined critical parameters are examined to optimize protein stability. Which variables are most important depends of the formulation type chosen. For example, resistance to agitation and/or accidental freezing is a critical property for an aqueous formulation, which would not be a concern for a lyophilized formulation. Similarly, choosing excipients that provide a glassy matrix to stabilize the protein is only a concern for dried formulations.

Formulation Options for Protein Pharmaceuticals

Different types of formulations need to be developed based on clinical needs, patient compliance, delivery method, stability of the drug, storage and distribution, and market competitiveness. Having a clear plan about what type of formulation is desired will allow one to design better formulation studies. Liquid formulations have been generally preferred due to the convenience of manufacturing and use. However, protein drugs may not be stable enough to be handled as a liquid formulation. Dried formulations (e.g., lyophilized) or suspension formulations (e.g., insulin zinc suspension) have been successfully used to overcome stability problems. In addition, specific applications and delivery may demand the appropriate type of formulation, e.g., a spray-dried powder for pulmonary delivery. As details of the stability issues are discussed in other chapters in this book, only practical issues regarding these different types of formulations will be summarized here.

Liquid Formulations. It is important to understand that developing conditions to keep proteins stable in a liquid form for a pharmaceutically relevant storage time (e.g., two years) is not a simple task. For most proteins, including relatively stable ones, at least some degradation should be expected even during refrigerated storage. Unless there is strong evidence supporting that the protein remains stable for two years and the evidence is supported by a broad spectrum of analytical methods, one needs to be very careful about the final decision to market a liquid formulation. Physical or chemical changes that have low activation energy should be monitored especially carefully under real storage conditions. If there is a degradation product that can significantly affect the pharmaceutical quality at its minimum concentrations (e.g., formation of particulates) great care must be taken to assure that such a product does reach unacceptable levels during real-time storage. If long-term storage studies are not implemented until late in the product development process, there is a risk that problems due to unacceptable levels of degradation products will not be discovered in time to test alternative formulations. In addition to stability during storage, temporary exposure to temperatures outside the recommended conditions should also be tested because it can affect the quality of the drug. Also, results obtained from multiple freeze-thawing cycles will be useful to determine if the formulation can accommodate unexpected freezing during distribution and storage. Likewise, sensitivity to agitation or surface denaturation needs to be understood to support formulation choices as well as shipping and handling guidelines. For formulations containing low protein concentration, special attention is required to avoid impurity-related stability issues (e.g., oxidation fostered by metal contaminants in excipients).

Solid Dosage Forms. Recent improvements in devices designed for easier use of lyophilized products, e.g., dual chamber syringes, dual chamber cartridges and convenient reconstitution devices, have helped pharmaceutical industries to develop lyophilized products without too many concerns surrounding patient compliance issues. Because lyophilized products have less stability-related issues and have a much greater potential tolerance for room-temperature storage, many biopharmaceutical industries consider the lyophilized formulation as a default option. In fact, lyophilization has been widely utilized to overcome various stability issues of labile proteins. However, it is important to note that lyophilization can present its own challenges, particularly in designing appropriate formulations and economic cycles. Detailed discussion about the rational design of stable lyophilized formulations is available in another chapter in this book.

In addition to the required stability, a successful lyophilized formulation should also have the desired physical properties of the dried powder, e.g., ruggedness of the cake and the maintenance of the physical states for each of the ingredients. For example, crystallization of an initially amorphous excipient during storage can lead to unacceptable degradation of the protein product and loss of cake structure. Furthermore, optimal water content for protein stability should be studied and controlled during storage. It is preferable to define an acceptable range of water contents to provide flexibility in the manufacturing process. To assure that the appropriate water content is maintain, the integrity of container/packaging should be monitored in a high relative humidity environment.

The identity and volume of the reconstitution medium should also be clearly defined, because the use of injectable fluids other than the recommended one may compromise the quality of the product (e.g., solution tonicity or protein stability). Reconstitution should be convenient and rapid, and the stability of the reconstituted solution during handling and delivery needs to be demonstrated.

Only recently have spray-dried formulations become commonly used with the advent of novel protein delivery technologies. For example, spray-dried powders have been useful for developing products that require uniform particle size (e.g., pulmonary inhalation for systemic delivery). Similar issues to the lyophilized formulation apply to the spray-dried formulation, and the details of development of spray-dried formulations are presented in another chapter in this book. In addition to protein stability, physical properties pertinent to powder processing and powder stability (e.g., flowablity, hygroscopicity, agglomeration, density and crystallization of excipients) need to be well characterized.

Single Dose and Multidose Forms. Most protein pharmaceuticals are marketed as single dose forms. However, multidose formulations are useful when the dose needs to be split (e.g., dose titration or dose combination). Multidose formulations are distinguished by the presence of preservatives in the formulation, which prevent microbial contamination and/or growth during multiple disruptions of container closure integrity. In general, the addition of preservative(s), regardless of the preservative used, significantly changes the stability profiles of proteins (Maa and Hsu, 1996; Fransson et al., 1997; Lam et al., 1997). In some extreme cases, visible precipitation and aggregation have been reported. Therefore, the effect of various preservatives on the stability of protein should be carefully examined. Other experiments required to qualify a multidose formulations include the preservative effectiveness test and the stopper self-sealing test. Detailed procedures and specification are available from the USP (U.S. Pharmacopeia), the Ph. Eur. (European Pharmacopeia) and the JP (Japan Pharmacopeia).

Desired properties of the ideal preservative include: effectiveness at low concentration against a wide variety of organisms; chemical stability; solubility; compatibility with the protein drug, excipients and auxiliary agents; free from objectionable odor, taste, color and stinging; and non-toxic and non-sensitizing both internally and externally at the required concentration. Also, it must not absorb, penetrate, or interact with containers or closures (Thompson, 1998). For further details on preservatives, readers are referred to Regulatory Document 4 in the Appendix and Thompson (1988). Examples of typical preservatives used for parenteral protein pharmaceuticals are benzyl alcohol, phenol, *m*-cresol and benzalkonium chloride.

Typical Protein Stability Problems: Causes and Solutions

Table 6 summarizes typical stability problems observed during protein formulation development and potential methods to solve each problem. The list does not represent the complexity of multiple problems that can be experienced with a given protein. Formulation research should be designed to handle each protein based on its unique stability profile.

Optimization of Formulation Variables

Overview of the Process. The optimization of formulation variables for product stability is the most critical part of protein formulation development. Various formulation excipients and buffers (Table 7) can be utilized and must, therefore, be chosen to maximize the pharmaceutical quality of the product (i.e., stability and activity) without introducing significant side effects. Among the listed formulation variables, the most powerful one is pH. Problems associated with the physical properties of a protein, e.g., precipitation due to solubility and/or stability, are generally very difficult to manage by other formulation means. Optimization of pH is a simple but very useful solution for such problems (Kolvenbach et al., 1997). Most chemical reactions also are affected by pH, e.g.,

Table 6.
Typical Stability Problems Observed in Protein Pharmaceuticals

Problems	Potential causes	Possible solutions
Non-covalent aggregation	Solubility, structural changes, heat, shear, surface, denaturants, impurities	pH, ionic additives, amino acids, surfactants, protein concentration, raw material purity
Covalent aggregation	Disulfide scrambling, other unknown mechanisms	pH, inhibit non-covalent aggregation
Deamidation	pH < 5.0 or pH > 6.0	pH optimization
Cyclic imide	pH around 5	pH optimization
Cleavages	Protease impurity, other unknown mechanisms	pH, product purity, inhibitors
Oxidation	Active oxygen species, free radicals, metals, light, impurity	Excipient purity, free-radical scavenger, active oxygen scavengers
Surface denaturation, adsorption	Low protein concentration, specific affinity, protein hydrophobicity	Surfactants, protein concentration, pH

Table 7.
Important Components of Protein Formulations

Formulation Variables	Desired attributes	Examples
pH	Provides good physical properties of protein, minimize degradations	
Stabilizer	Inhibit degradations, effective at low concentrations	Surfactants, sugars, salts, antioxidants
Solubilizer	Improve the solubility, effective at low concentrations	Salts, amino acids, surfactants
Buffer	Good buffering capacity, stable to temperature change, stable to freezing, good safety record	Phosphate, acetate, histidine, glutamate
Tonicity modifier; bulking agent	Inert, good safety record	Sodium chloride, sorbitol, mannitol, glycine

deamidation, cyclic imide formation, disulfide scrambling, peptide bond cleavage, and oxidation (see reviews listed above). Other functional excipients should be also carefully evaluated for the benefit of the product (e.g., use of sucrose to stabilize protein during lyophilization and storage in the dried solid).

In addition to their intended use, excipient candidates should be also qualified as appropriate pharmaceutical ingredients. Generally, it is preferred to select excipients that have been used in marketed products with a relevant route of delivery. Further, it is preferable if these excipients have been used with similar frequency of dosing, history of chronic use and similar patient populations. Otherwise, the approval and safety of the excipients need to be carefully examined. If the excipient is considered safe with a solid scientific basis or has a proven clinical safety record, it can be considered equivalent to approved excipients. The list of excipients used for parenteral pharmaceuticals is available in the literature (Powell et al., 1998; Nema et al., 1997; Appendix Regulatory Document 5). When it is necessary to introduce other excipients with minimum safety records, a significant risk associated with the excipients will be added to the product development and additional pre-clinical and clinical studies may be needed.

Another important requirement in qualifying an excipient is the purity of the raw material. Depending on the historical use as a pharmaceutical ingredient, several different pharmaceutical grades are available, e.g., USP (U.S. Pharmacopeia), Ph. Eur. (European Pharmacopeia) and JP (Japan Pharmacopeia). These pharmaceutical grade materials should be considered as a primary resource, but the quality provided may not be good enough for specific product development. For example, significant stability problems can be found with some impurities even at concentrations below their specification, e.g., metal ions, peroxides, proteases and reducing sugars. These problems are more prominent in low protein concentration products due to a high impurity-protein ratio, although problems like visible precipitation of the protein may be independent of protein concentration. If adjustment in the existing specification is necessary, it is critical to look into the availability of GMP quality raw materials with modified specifications as early as possible, once the potential problem is identified.

The use of excipients derived from animals (e.g., Tweens) or humans (e.g., human serum albumin) should be avoided if possible due to the risk associated with transmissible diseases like bovine spongiform encephalopathy, Creutzfeldt-Jakob Disease, hepatitis virus and HIV. Numerous regulatory guidelines have been issued to discourage the use of animal/human-derived excipients (Appendix Regulatory Documents 6,7,8). When animal-derived excipients have to be included in the product, the manufacturer will need to demonstrate that their selection is fully justified.

Design of the Study. A design for the formulation optimization study should be in place before commencing the work. A typical study protocol will include the following information:

- Study title
- Study objective
- Source and quality of drug substance and excipients
- Material preparation
- Formulation matrix
  - Formulation variables
  - Protein concentrations or bracket
- Analytical methods
- Storage conditions (temperature, light, humidity)
- Additional sample handling conditions (temporary exposure to stresses, container orientation, etc.)
- Sampling schedule and expected duration of the study
- Plan for data analysis and report

The study can be more efficiently designed by utilizing experimental design software packages, which will help to minimize the resources required for sample preparation, analyses, and data analysis.

Necessary Studies for Formulation Development

Storage Stability Study. Documenting that the formulation will keep the protein stable until the desired expiry can be the most time-consuming part of formulation development. The expiry requirements are determined by distribution not regulatory requirements. One year is probably too short for effective manufacturing and distribution through normal channels. In general, a shelf life of 18 month is considered acceptable for commercialization. Results obtained from accelerated stability studies are useful for predicting potential degradation products and appropriate analytical methods, but whatever the actual shelf life, it must be supported by sufficient real-time storage data to obtain regulatory approval. Thus, it is important to establish a final formulation and start the real-time storage studies as early as possible during product development.

Process Development. Formulations that can be prepared on a small scale without experiencing any problems may encounter significant problems during the scale-up of the process. For example, mixing solutions in a large stainless steel tank, pumping solutions through stainless steel tubing, filtration and filling through a high-speed filling machine can introduce unexpected stresses to the protein. An increase in the formation of particulates, along with a loss of proteins due to surface adsorption and aggregation, has routinely been observed. It is important to expose the formulation to equivalent stresses and make sure that no formulation adjustment is necessary to accommodate the manufacturing processes. Again, this testing should be done as early in the development process as possible.

Transportation, Handling and Delivery Study. Unexpected environmental changes can be encountered during the distribution and handling of products (e.g., accidental freezing, exposure to temperatures different from the recommended conditions, vigorous agitation, etc). It is critical to develop the formulation with these stresses in mind because they can compromise the quality of the product.

Also, during administration to the patient, proteins can be exposed to different types of stresses introduced by the device and the routes of delivery. Examples include incompatibility with the delivery device (e.g., protein aggregation induced by exposure to tubing surfaces) and/or concomitant medication (e.g., protein aggregation induced by co-administered antibiotics).

Preclinical and Clinical Studies. Results documenting the maintenance of the biophysical and biochemical properties of the protein are essential for a final formulation decision. Before finalizing the formulation, it is also important to confirm that it does not affect critical *in vivo* biological properties of the protein (e.g., activity, pharmacokinetic profile, and toxicity profile). Maintenance of a protein's biophysical properties can be examined using various structural analyses (e.g., circular dichroism, fluorescence and infrared spectroscopies, etc.). It is possible to determine the biochemical equivalence of the protein pharmaceutical by *in vitro* activity and/or preclinical *in vivo* bioassays. Results supporting the toxicity profile of the formulation can be generated by both preclinical and clinical studies.

Strategies to Overcome Difficult Formulation Problems

Occasionally, problems are encountered that are difficult to overcome with conventional formulation approaches. It is possible to conduct additional studies to confirm that the problem is pharmaceutically acceptable. On the other hand, unique formulation approaches can be introduced to address specific problems. Examples of this approach are provided below.

Qualification of Degradation Products. If decreasing the amount of degradation product below the specified threshold is not feasible, then the degradation product can be defined as a drug substance by demonstrating that it does not affect the pharmaceutical quality of the drug product, i.e., safety and efficacy (Appendix, Regulatory Document 1). If there is not sufficient information available, additional qualification studies recommended in the regulatory guidance can be carried out (see the discussion above). The experience of having these degradation products in the materials used for clinical studies can sometimes be very useful because the results obtained can provide useful insight into the clinical implication of the degradation products.

Site-directed Mutagenesis to Improve Properties. Some problems related to the intrinsic properties of a protein cannot be overcome unless a change in the sequence is introduced. After careful research is carried out to identify the problematic region or residue, the protein can be engineered for better physical and/or chemical properties. For serious problems like precipitation due to insolubility or poor stability, changes in protein sequence have been proven effective for improving physical properties (Murby et al., 1995; Roig and Kennedy, 1995).

Chemical Modifications. The physical properties of proteins can be improved by modifying problematic amino acid side chains with small compounds or large polymers (e.g., attachment of polyethylene glycol) (Fagain, 1995; Guerra et al., 1998; Francis et al., 1998). In addition, there are other desirable improvements that can be achieved by the conjugation chemistry (e.g., enhanced pharmacokinetic profiles, reduced immunogenecity, enhanced adsorption, etc.).

Unconventional Dosage Forms. Some protein stability problems can be resolved by introducing novel approaches in the formulation. Examples include suspension formulations (Defelippis et al., 1998), microencapsulation with cyclodextrin (Brewster et al., 1991), suspension of dry-powder in non-aqueous vehicle (Knepp et al., 1998), and use of non-aqueous vehicles with hydrophobic ion-pairing (Manning et al., 1995).

FORMULATION IN COMMERCIAL PRODUCT DEVELOPMENT

Critical Formulation Decisions During Pharmaceutical Development

A brief summary of the important phases of a commercial product development is shown in Figure 1. Ideally, a stable formulation would be available during the early discovery research period to ensure that the protein is in its active and stable form during critical feasibility studies. However, development of a stable formulation for each drug candidate is not practically possible because such an approach would take too much time and too many resources. Generally, formulation development starts after the decision is made to start clinical trials. A

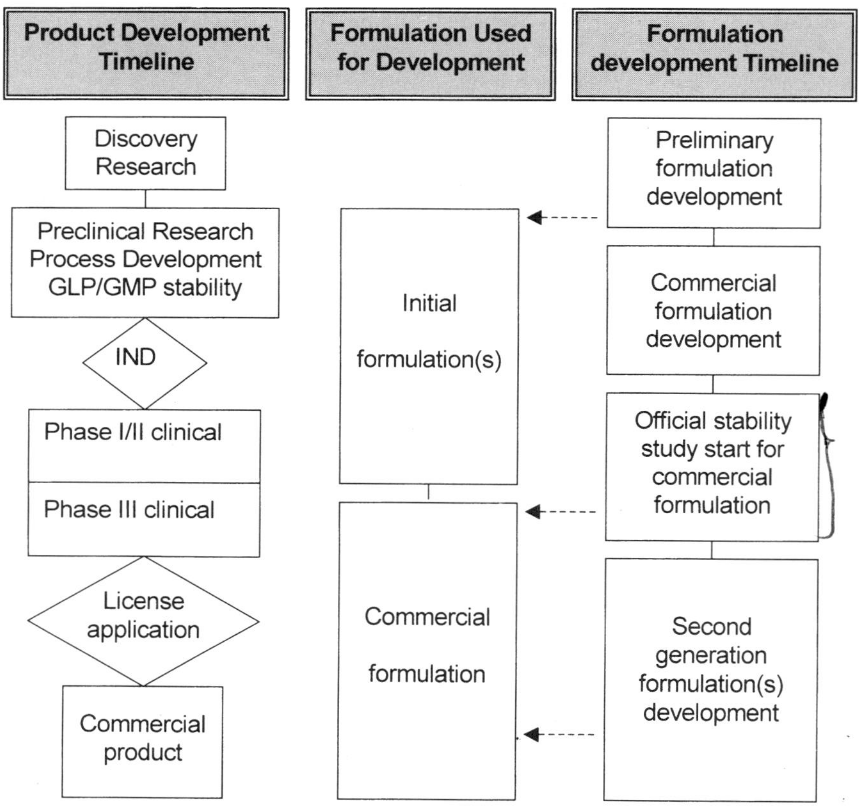

Figure 1. Diagram showing commercial formulation development process.

reliable formulation is required to support various preclinical and clinical studies. The formulation can be further improved later to satisfy the clinical, marketing, and regulatory needs. Although formulation development does not necessarily appear rate-limiting during this improvement period, it is important to understand that changing a formulation may require much more supporting work, including additional clinical trials, which can take years.

Formulation for Early Preclinical and Clinical Studies

When a promising protein drug candidate is identified from preliminary feasibility studies, a decision will be made to introduce the drug to clinical studies. In order to obtain regulatory approval to initiate the clinical trials, it is required to demonstrate safety, manufacturing capability, stability of the drug and reproducibility in studying the drug. A sufficiently stable formulation is necessary at this stage because it is important to maintain the quantity and quality of the protein for all research studies from which results will be used for registration of the product with regulatory agencies. Therefore, formulation development becomes a rate-limiting process at this early stage. Formulations can be stored frozen or lyophilized for this purpose. At this stage in product development, the shelf-life requirement is determined by the logistics of supplying drug for clinical trials.

Commercial Formulation

Commercially viable and market competitive formulations have some common features. Most of all, the formulation should maintain the safety and efficacy profile of the protein drug during all the handling and uses specified on the label. Since commercial distribution channels are not equipped for frozen products, shipping and storage at refrigerated temperature or higher are required. Sufficient shelf life needs to be determined under conditions to which the product will be exposed in the commercial distribution system. Various systematic studies need to be carried out to comply with regulatory requirements for registration. It takes 1–2 years to collect all of these results, so the commercial formulation is developed while preclinical studies or early clinical trials (with the preliminary formulation) are in progress. If possible, the commercial formulation should be introduced before the pivotal clinical trial because clinical experience is the most effective way to confirm the safety and efficacy aspects of the formulation. In addition, formulation changes after this point may introduce formidable challenges to the clinical program and to obtaining regulatory approval.

Regulatory Issues in Formulation Development

Incorporating regulatory guidelines into the formulation development is not only useful to develop high quality pharmaceutical formulations for maximum benefit to the patients, but also critical to prepare complete documents necessary for regulatory approval for commercialization of the drug. Readers who wish to obtain a comprehensive understanding of the guidelines are referred to the relevant documents published by regulatory agencies. In this section, a brief summary of the guidelines related to the formulation issues is provided.

Guidelines for Stability Studies. Comprehensive guidelines can be obtained from regulatory agencies (Appendix, Regulatory Documents 9–12) and review articles in the literature (Grimm, 1998; Kommanaboyina and Rhodes, 1999; Matthews, 1999). These guidelines generally cover the formal stability studies for bulk drug material, in-process samples and final formulated product. Although some issues may not be relevant for the earlier formulation development work, these guidelines provide key elements for designing stability studies. The guidelines provide information such as what types of experiments should be included in the study protocol, how to propose a stability-indicating profile, and what analytical results are needed to define the purity of the product and the molecular characteristics of degradation products. Another important piece of information provided in the guidelines is the definition of proper experimental conditions, e.g., definition of storage temperature, humidity, light strength, and accelerated and stressed conditions. The guidelines also provide the official protocols to evaluate the results obtained from stability studies.

Results Required to Apply for Regulatory License for a Drug Product. Detailed information regarding the formulation is presented at the Development Pharmaceutics section in the CMC section of regulatory applications (Appendix, Regulatory Docoments 13–16). All of the essential information required to apply for regulatory approval of protein formulations can be found in Note for Guidance on Development Pharmaceutics (Appendix Regulatory Document 17) and Development of Pharmaceutics for Biotechnological and Biological Products (Appendix, Regulatory Document 8). It is important to note that this is currently a European requirement although the international harmonization process may eventually introduce it to U.S. regulatory requirements. A brief summary of the important formulation information is shown in Table 8.

Results Required to File Formulation Amendments. A formulation change can introduce substantial potential adverse effects on the identity, strength, quality, purity or potency of the product as related to the safety or effectiveness of the product. For this reason, it is required that a supplement to the approved license application be submitted. This supplement should include a detailed description of the proposed change. It should also include methods and results for the studies performed to evaluate the effect of the change on the

Table 8.
Summary of Information Included in Regulatory Applications

	General information	Comments specific to proteins
Active substance (protein)	Compatibility with excipients and other combined products, and physicochemical characteristics	Structural elements responsible for the biological activity; good coverage of degradations occurring during storage, manufacturing, and delivery.
Excipients	Function of each excipients, justification for their inclusion, and compatibility with other excipients, choice of quality	
Formulated products	Overage, physicochemical parameters, components with appropriate results supporting intended purpose; compatibility with diluent, device, and other drugs in contact before delivery; critical physical properties	Stability of structure in terms of biological activity; formulation optimization for both manufacturing process and stability
Packaging material	Integrity of container and closure during storage, reconstitution, admixture, dilution; sorption to container, leaching, and dose reproducibility	Adsorption, denaturation at the interface, and aggregation of the surface-denatured protein; reconstitution of dry-powder formulation
Manufacturing process	Manufacturing process for the preparation of formulation and its justification, appropriate method of sterilization and justification	Results to support the quality and stability of the protein during the manufacturing process; membrane filtration under aseptic conditions sufficient

product's identity, strength, quality, purity, and potency of the product, as related to the product's safety or effectiveness (Appendix, Regulatory Documents 18–20). When the degradation profile is changed qualitatively or quantitatively, it is recommended to follow the impurity-related guidelines discussed above (Appendix, Regulatory Documents 2,3). The manufacturer must obtain approval of the supplement by FDA prior to distribution of the product made using the change. In general, the following studies (Table 9) can be carried out to confirm

Table 9.
Studies Needed to Support Change in Formulation

Research requirements	Supporting results
Purity	Stability, compatibility, Structural analyses
Potency	*In vitro/in vivo* bioassays, preclinical and/or clinical pharmacokinetic comparability , clinical efficacy
Safety	Preclinical safety, clinical safety

that the change in the formulation will not affect the safety and effectiveness of the product. It is recommended to consult the regulatory representative to find out how much additional information may be needed for the approval of the changed formulation.

APPENDIX: LIST OF REGULATORY DOCUMENTS

1. International Conference on Harmonization: Guidance on specifications: Test procedures and acceptance criteria for Biotechnological/Biological Products. Federal Register: August 18, Volume 64, Number 159, pp. 44928–44935 (1999).
2. International Conference on Harmonization: Impurities in New Drug Products. FDA Q3B (11/97).
3. Guidance for Industry. ANDAs: Impurities in Drug Products. Draft guidance. FDA (12/98).
4. Note for Guidance on Inclusion of Antioxidants and Antimicrobial Preservatives in Medicinal Products (CPMP/CVMP/QWP/115/95) EMEA (7/97).
5. Inactive ingredient guide: inactive ingredients for currently marketed drug products. (1996) FOI Services, Inc. Rockville, MD.
6. Note for guidance on minimizing the risk of transmitting animal spongiform encephalopathy agents via medicinal products. EMEA (4/99).
7. Note for guidance on plasma-derived medicinal products. EMEA (7/98).
8. Development Pharmaceutics for Biotechnological and Biological Products. Annex to Note for Guidance on Development Pharmaceutics (CPMP/QWP/155/96). EMEA (10/1999).
9. International Conference on Harmonization: Stability testing of new drug substances and products. Federal Register, Sept. 22, Volume 59, Number 183, pp. 48754–48759 (1994).
10. International Conference on Harmonization: Final guideline on stability testing of biotechnological/biological products. Federal Register, July 10, Volume 61, Number133, pp. 36466–36469 (1996).
11. Guideline for Industry: Stability testing for drug substances and drug products: draft guidances. FDA, (6/98).
12. International Conference on Harmonization: Guidelines for the photostability testing of new drug substances and products. Federal Register, May 16, Volume 62, Number. 95, pp. 27115–2712 (1997).
13. Guideline for Submitting Documentation for the Stability of Human Drugs and Biologics FDA. (2/87).
14. Guidance for Industry. For the Submission of Chemistry, Manufacturing and Controls and Establishment Description Information for Human Plasma-

Derived Biological Products, Animal Plasma or Serum-Derived Products. FDA (2/99).
15. Guidance for Industry. INDs for Phase 2 and 3 Studies of Drugs, Including Specified Therapeutic Biotechnology-Derived Products Chemistry, Manufacturing, and Controls Content and Format (Draft guidance) FDA (2/99).
16. Content and Format of Investigational New Drug Applications (INDs) for Phase 1 Studies of Drugs, Including Well-Characterized, Therapeutic, Biotechnology-Derived Products, FDA (11/95).
17. Note for Guidance on Development of Pharmaceutics (CPMP/BWP/328/99). EMEA (01/1998).
18. Demonstration of Comparability of Human Biological Products, Including Therapeutic Biotechnology-Derived Products, April, 1996.
19. Guidance for Industry: Changes to an Approved Application. For Specified Biotechnology and Specified Synthetic Biological Products, 21 CFR 601.12, 314.70; July 24, 1997, Vol 62. No. 142.
20. Guidance for Industry: Changes to an Approved Application: Specified Biotechnology and Specified Synthetic Biological Products; July 1997.

REFERENCES

Ahern, T.J. and Manning, M.C., 1992a. Stability of protein pharmaceuticals, Part A: Chemical and physical pathways of protein degradation. *Pharm. Biotech. Ser.* Volume 2. Plenum Press, N.Y.

Ahern, T.J. and Manning, M.C., 1992b. Stability of protein pharmaceuticals, Part B: In vivo pathways of degradation and strategies for protein stabilization. *Pharm. Biotech. Ser.* Volume 3. Plenum Press, N.Y.

Arakawa, T., Prestrelski, S., Kinney, W., and Carpenter, J.F., 1993. Factors affecting short-term and long-term stabilities of proteins. *Adv. Drug Delivery Rev.* 10:1.

Brewster, M.E., Hora, M.S., Simpkins, J.W., and Bodor, N., 1991. Use of 2-hydroxypropyl-beta-cyclodextrin as a solubilizing and stabilizing excipient for protein drugs. *Pharm. Res.* 8:792.

Cleland, J.L., Powell, M.F., and Shire, S.J., 1993. The development of stable protein formulations—A close look at protein aggregation, deamidation and oxidation. *Crit. Rev. Ther. Drug* 11:60.

Chen, T., 1992. Formulation concerns of protein drugs. *Drug Dev. Ind. Pharmacy*, 18:1311.

Defelippis, M.R., Bakaysa, D.L., Bell, M.A., Heady, M.A., Li, S., Pye, S., Youngman, K.M., Radzuik, J., and Frank, B.H., 1998. Preparation and characterization of a cocrystalline suspension of [Lys(B28),Pro(B29)] human insulin analogue. *J. Pharm. Sci.* 87:170.

Fagain, C.O., 1995. Understanding and increasing protein stability. *Biochimica et Biophysica Acta.* 1252:1.

Francis, G.E., Fisher, D., Delgado, C., Malik, F., Gardiner, A., and Neale, D., 1998. PEGylation of cytokines and other therapeutic proteins and peptides: the importance of biological optimisation of coupling techniques *Int. J. Hematology.* 68:1.

Fransson, J., Hallen, D., and Florin-Robertsson, E., 1997. Solvent effects on the solubility and physical stability of human Insulin-like Growth Factor I. *Pharm. Res.* 14:606.

Grimm, W., 1998. Extension of the international conference on harmonization tripatic guieline for stability testing of new drug substances and products to countries of climatic zones III and IV. *Drug Dev. Indust. Pharm.* 24:313.

Guerra, P.I., Acklin, C., Kosky, A.A., Davis, J.M., Treuheit, M.J., and Brems, D.N., 1998. PEGylation prevents the N-terminal degradation of megakaryocyte growth and development factor *Pharm. Res.* 15:1822.

Herron, J.N., Jiskoot, W., and Crommelin, D.J.A., 1995. Physical methods to characterize pharmaceutical proteins. *Pharm. Biotech. Ser.* Volume 7. Plenum Press, N.Y.

Jones, A.J.S., 1993. Analysis of polypeptides and proteins *Adv. Drug Del. Rev.* 10:29.

Knepp, V.M., Muchnik, A., Oldmark, S., and Kalashnikova, L., 1998. Stability of nonaqueous suspension formulations of plasma derived factor IX and recombinant human alpha interferon at elevated temperatures. *Pharm. Res.* 15:1090.

Kolvenbach, C.G., Narhi, L.O., Philo, J.S., Li, T., Zhang, M., and Arakawa, T., 1997. Granulocyte-colony stimulating factor maintains a thermally stable, compact, partially folded structure at pH 2 *J. Pept. Res.* 50:310.

Kommanaboyina, B. and Rhodes, C.T., 1999. Trends in stability testing with emphasis on stability during distribution and storage. *Drug Dev. Indust. Pharm.* 25:857.

Lam, X.M., Patapoff, T.W., and Nguyen, T.H., 1997. The effect of benzyl alcohol on recombinant human interferon-gamma *Pharm. Res.* 14:725.

Maa,Y.F. and Hsu, C.C., 1996. Aggregation of recombinant human growth hormone induced by phenolic compounds *Int. J. Pharm.* 140:155.

Manning, M.C., Matsuura, J.E., Kendrick, B.S., Meyer, J.D., Dormish, J.J., Vrkljan, M., Ruth, J.R., Carpenter, J.F., and Shefter, E., 1995. Approaches for increasing the solution stability of proteins *Biotech. Bioeng.* 48:506.

Manning, M.C., Patel, K., and Borchardt, R.T., 1989. Stability of protein pharmaceuticals. *Pharm. Res.* 6:903.

Matthews, B.R., 1999. Regulatory aspects of stability testing in Europe. *Drug Dev. Indust. Pharm.* 25:831.

Murby, M., Samuelsson, E., Nguyen, T.N., Mignard, L., Power, U., Binz, H., Uhlen, M., and Stahl, S., 1995. Hydrophobility engineering to increase solubility and stability of a recombinant protein from respiratory syncytial virus. *Eur. J. Biochem.* 230:38.

Nema, S., Washkuhn, R.J., and Brendel, R.J., 1997. Excipients and their use in injectable products *PDA J. Pharm. Sci. Technol.* 51:166.

Note for Guidance on Inclusion of Antioxidants and Antimicrobial Preservatives in Medicinal Products, 1997. (CPMP/CVMP/QWP/115/95) EMEA (7/97).

Pearlman, R. and Wang, Y.J., 1996. Formulation, characterization, and stability of protein drugs: case histories. *Pharm. Biotech. Ser.* Volume 9. Plenum Press, N.Y.

Powell, M.F., Nguyen, T., and Baloian, L., 1998. Compendium of excipients for parenteral formulations. *PDA J. Pharm. Sci. Technol.* 52:238.

Reubsaet, J.L.E., Beijnen, J.H., Bult, A., Van-Maanen, R.J., Marchal, J.A.D., and Underberg, W.J.M., 1998. Analytical techniques used to study the degradation of proteins and peptides: chemical instability *J. Pharm. Biomed. Anal.* 17:955.

Roig, M.G. and Kennedy, J.F., 1995. Perspectives for biophysicochemical modifications of enzymes. *J. Biomaterials Sci. Polymer Ed.* 7:1.

Thompson, J.E., 1998. Practical Guide to Contemporary Pharmacy Practice. Lippincott Williams & Wilkins, Hagerstown, MD.

Volkin, D.B., Mach, H., and Middaugh, C.R., 1997. Degradative covalent reactions important to protein Stability. *Molec. Biotech.* 8:5.

Wang, Y.J. and Hanson, M.A., 1988. Parenteral formulations of proteins and peptides: stability and stabilizers. *J. Parent. Sci. Technol.* 42:SS4.

Wang, Y.J. and Pearlman, R., 1993. Stability and characterization of protein and peptide drugs: case histories. *Pharm. Biotech. Ser.* Volume 5. Plenum Press, N.Y.

2

Recombinant Production of Native Proteins from *Escherichia coli*

Tsutomu Arakawa, Tiansheng Li, and Linda O. Narhi

INTRODUCTION

The production of large quantities of proteins became possible with the advent of recombinant DNA technology, and the subsequent expression of recombinant proteins in *Escherichia coli* (*E. coli*) (Itakura, 1977). Recombinant proteins are produced in *E. coli* cytoplasm at high concentrations in a very different environment than that in which they are normally expressed. Their structure, stability and solubility are affected by this environment and can be quite different than that of the native protein. Therefore, purification of recombinant proteins often requires different techniques than those used to purify the same protein from its natural environment. Numerous attempts have been made to purify and/or refold expressed proteins.

Heterologous expression of recombinant proteins in *E. coli* results in insoluble inclusion bodies, soluble proteins, or a distribution of the proteins between the insoluble and soluble fractions. Proteins expressed as insoluble inclusion bodies must be solubilized, purified and refolded. If they contain cysteines forming intra- and/or inter-chain disulfide bonds, oxidation of free cysteines is

Tsutomu Arakawa • Alliance Protein Laboratories 3957 Corte Cancion, Thousand Oaks, CA 91360. *Tiansheng Li and Linda O. Narhi* • Amgen, Inc. Thousand Oaks, CA 91320.

Rational Design of Stable Protein Formulations, edited by Carpenter and Manning. Kluwer Academic / Plenum Publishers, New York, 2002.

required during the refolding process. Expressed proteins located in the soluble fraction of the bacteria usually are those that contain no disulfide bonds in the native state. However, there are some exceptions to this rule. Even disulfide-containing proteins can be expressed in the folded conformation or in a soluble, but misfolded, conformation. When the expressed proteins are already folded, their distribution in the supernatant should be maximized and the supernatant used as the source for purification. When the expressed proteins are soluble but misfolded, they need to be denatured and refolded using a procedure similar to that used for insoluble inclusion bodies. This paper will review the purification and refolding procedures that have been applied to insoluble inclusion bodies and to proteins expressed in *E. coli* as soluble folded or misfolded forms.

DISTRIBUTION OF EXPRESSED PROTEINS

Depending on the absence or presence of an N-terminal signal sequence, the expression of heterologous protein in *E. coli* can be directed into cytoplasmic or periplasmic space, as summarized in Figure 1. All outer membrane and periplasmic proteins in *E. coli* are synthesized with a signal peptide and secreted through the cytoplasmic membrane export machinery (Michaelis and Beckwith, 1982; Pugsley, 1993; Wickner et al., 1991). Any signal peptides that are used by the secretion machinery can be used to direct the heterologous proteins into the periplasmic space (Chan et al., 1981; Talmadge et al., 1980a, 1980b; Chang et al., 1983; Ghrayeb et al., 1984; Collmer et al., 1985; Takahara et al., 1985; Duffaud et al., 1987; Skerra and Plückthun, 1987; Better et al., 1988). Successful application of this strategy usually occurs when the foreign proteins that are fused to the signal sequence are also originally secretory proteins.

Some *E. coli* or foreign proteins exported through this machinery are known to be secreted into the culture media across the outer membrane (Mackman and Holland, 1984; Müller et al., 1983; Kudo et al., 1983; Abrahmsen et al., 1986; Horuk et al., 1997; Nagahari et al., 1985), as shown in Figure 1. Although the precise mediator of secretion of proteins into the culture media is not known, the outer membrane of *E. coli* possesses the OmpA and OmpG proteins, which can serve as passive diffusion pores for small hydrophilic molecules (Nikaido and Vaara, 1985). Therefore, the hydrodynamic size and hydrophilicity of proteins secreted into the periplasmic space appear to play critical roles in determining the distribution of the secreted protein between the culture media and the periplasmic space.

Successful application of this periplasmic secretion strategy results in the expression of a correctly cleaved and folded protein. For example, periplasmic secretion of anthrax toxin antigen was made with the OmpA signal, resulting in

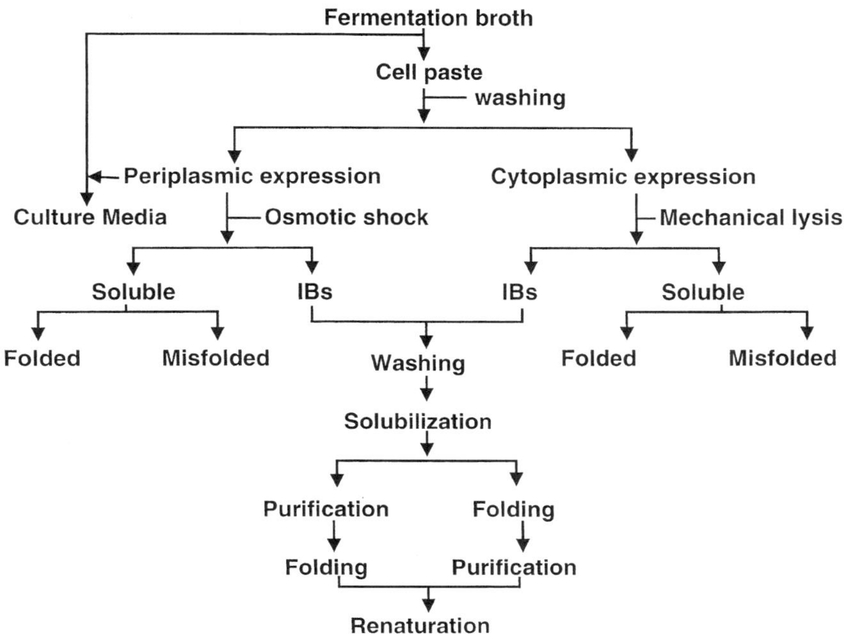

Figure 1. Distribution of recombinant proteins expressed in *E. coli.* IB's: inclusion bodies.

formation of a soluble mature protein (Sharma et al., 1996). However, a number of exceptions have been reported. Erythropoietin binding protein was directed to the periplasm as either a pelB or OmpT signal sequence. Both constructs resulted in insoluble pellet, with the pelB signal partially retained (Johnson et al., 1996). Granulocyte macrophage colony stimulating factor (GM-CSF) was expressed as an insoluble protein using the OmpA secretion system (Libby et al., 1987). An even more complicated case is that of human interleukin-1β (IL-1β) (Denefle et al., 1989; Jung et al., 1988). This protein can be expressed in the cytoplasm in a folded, active form with the signal peptide correctly cleaved, and it is released from the cells by osmotic shock (Jung et al., 1988). However, when the protein was expressed as the PhoA fusion protein, processing did not occur, and the protein was located entirely in the insoluble cell pellet as the full-length tagged protein. When the OmpA-IL-1β construct was expressed, the processing was somewhat improved, with 50–100% of the protein existing as the mature IL-1β. However, the protein was still located entirely in the insoluble fraction. Fragments of IL-1β expressed as OmpA-fusion proteins were found to be 100%

processed, but the processed fragments were also localized in inclusion bodies. Fusion proteins made using IL-1β with the deletion of the fourteen C-terminal amino acids also resulted in constructs that were localized in inclusion bodies. As was seen for the full-length IL-1β, the PhoA fusion protein was not processed, but isolated as the full-length protein, while the OmpA fusion protein was isolated as the mature C-terminally cleaved IL-1β. When the standard protein OmpA-alkaline phosphatase was expressed at similar levels, it was localized as a mature, folded protein in the periplasmic space. Thus, the incomplete cleavage of the signal peptide, and the lack of transport into the periplasmic space for some of these constructs cannot be attributed to the level of expressed proteins overwhelming the processing machinery. Instead, it suggests that interactions can occur between the protein and the tag that interfere with expression as a soluble protein and that are protein (sequence) specific. This specificity depends on both the tag and the protein.

Because of its highly efficient secretion (Halfmann et al., 1993), the MalE-fusion system has been one of the most frequently employed choices for soluble fusion-protein expression. This system was successfully used to express the N-terminal portion of the cytokine receptor homology (CRH) domain of granulocyte colony stimulating factor (G-CSF) receptor (containing ~100 amino acids) (Hiraoka et al., 1995). Maltose-binding protein (MBP), with its own signal sequence, was fused to this construct. Its expression resulted in secretion of MBP-receptor fusion protein and, after processing, production of a receptor capable of high affinity ligand binding. Tanaka et al. (1998) expressed the entire CRH domain of G-CSF receptor using the PelB-fusion system and reported that the purification yield from the system has almost 100-fold higher than that from the MalE-fusion system. However, while the CRH domain was secreted into the periplasm, it was insoluble and required 6 M guanidine HCl for solubilization, followed by refolding and purification.

A factor as specific as the *E. coli* strain used can affect the success of periplasmic secretion. An Fc fragment of a human antibody was recently expressed in two different *E. coli* strains with the identical fusion protein cloned into the bacteria. In one case, secretion of soluble, folded, and active protein resulted. However, in the other case, soluble protein was isolated from the periplasm, but was found to be inactive, with differences from the native protein in both secondary and tertiary structure as detected by circular dichroism, fluorescence and infrared spectroscopies (Burgess, Wypych, Stevenson, Narhi and Jiang, unpublished results).

Direct expression of mature proteins, with an initiation Met residue as the N-terminus, results in either cytoplasmic soluble protein or insoluble inclusion bodies. Foreign proteins, such as intracellular enzymes, nuclear receptors, signaling regulators and their domains, all of which are cytoplasmic in their native environment, are generally expressed as soluble proteins. Two cytoplasmic enzymes, which were expressed as soluble proteins, were described by Yu et al. (1996) and Fletcher et al. (1996). The dihydrofolate reductase domain of *Leishmania major* dihydrofolate reductase-thymidylate synthase bifunctional protein was expressed as a soluble protein in *E. coli* strain PA414. The expressed protein was purified as a single folded protein from the bacterial lysate using MTX-sepharose affinity chromatography (Yu et al., 1996). HIV-1 recombinant reverse transcriptase was expressed as a soluble protein in E. coli strain JM109. This protein was expressed in the bacterial lysate, and purified by using a combination of Q- and S-columns (Fletcher et al., 1996).

Foreign proteins that are secretory proteins are usually expressed as insoluble proteins localized in inclusion bodies. However, some subsets of this class of proteins are efficiently accumulated as active, folded proteins.

To our knowledge, basic proteins without disulfide bonds, such as fibroblast growth factor (FGF) family members (Denefle et al., 1989; Jung et al., 1988) and interferon-γ (IFN-γ), tend to be expressed, at least in part, as soluble proteins. Distribution between soluble and insoluble fractions depends on the specific proteins and on the lysis conditions and solvents used, as well as the fermentation conditions. Even keratinocyte growth factor (KGF), which has three intra-chain disulfide bonds, can appear in both fractions. Distribution of expressed protein in the soluble fraction can be improved by carefully choosing the solvents used for cell lysis, etc. (described below). In addition, the level of soluble protein can be increased by manipulating cell growth conditions. For example, growth of the bacterial host at lower temperatures, rather than at 37°C, occasionally improves protein solubility (Wulfing and Plückthun, 1994; Schein, 1989), as does varying the length, conditions and timing of induction (Sawyer et al., 1994). Addition of a specific ligand for the protein can also increase the expression level of a soluble protein (Proudfoot et al., 1996). Some success has also been reported with the coexpression of chaperonins (Caspers, et al., 1994; Proudfoot et al., 1996) or thioredoxin (Yasukawa et al., 1995); the presence of these proteins at high concentration increased the expression level of soluble proteins. Proteins expressed in a soluble form may not necessarily be folded correctly, or active, in which case refolding is required. An example is agouti-related protein, which was found in the soluble fraction of *E. coli* as large, intermolecular disulfide-linked aggregates (Rosenfeld et al., 1998). Following denaturation, reduction, and then oxidation and refolding, an active, folded molecule was obtained. Another example of this behavior is a KGF receptor fusion protein. This protein was localized in the cytosol as very large, soluble aggregates consisting of DNA, and covalent disulfide and non-disulfide linked oligomers (M. Snavely, private communication).

CELL WASHING AND LYSIS

Except for rare occasions when proteins are expressed and secreted into the culture medium, *E. coli* cells must be chemically or mechanically lysed to release the expressed proteins. Protein localized in the periplasm can also be released by conventional osmotic shock (Koshland and Botstein, 1980). However, this procedure also results in the release of at least some cytoplasmic proteins (Denefle et al., 1989; Schilstra et al., 1984; Jacobson et al., 1976). When working with proteins expressed as soluble, active molecules, it is important to maximize the amount of the protein in the supernatant of lysed cells. Water is customarily used to lyse the cells. However, other buffers can be used to improve the recovery of the protein in the supernatant. It is important that the folded structure of the expressed proteins be maintained throughout this procedure. Hence, exposure to harsh conditions should be avoided. Harsh conditions include low pH (e.g., below 5), high pH (above 9), or exposure to strong detergent (e.g., SDS, deoxycholate), urea (e.g., above 2 M) or guanidine HCl (e.g., above 1 M). However, there are some proteins that are stable in low pH (Krensky and Nelson, 1997).

Many members of the FGF family have been purified from the supernatant of lysed *E. coli* cells, using water or dilute buffers to generate a supernatant containing a large amount of the protein expressed in soluble form (Arakawa et al., 1993; Fox et al., 1988; Hu et al., 1998). However, KGF (FGF-7) (Chen et al., 1994), FGF-10 (unpublished) and FGF-16 (Danilenko et al., 1999) required different handling than the other family members. FGF-7 (KGF) shows significant localization in the soluble fraction when lysed in 50 mM sodium phosphate containing 0.1 M NaCl, a buffer that has a relatively high ionic strength (Chen et al., 1994). FGF-16 is a novel member of the FGF family that was cloned using the conserved sequence between FGF-3 and FGF-7 (Miyake et al., 1998). When *E. coli* cells expressing this gene, with either the full-length sequence or the protein with the 34 N-terminal amino acids truncated, were lysed in water or dilute buffer, little protein was found in the soluble fraction. However, the protein concentration in this soluble fraction was greatly enhanced when the cells were broken in 1 M ammonium sulfate. These types of observations clearly demonstrate the importance of selecting optimal conditions for isolating recombinant proteins expressed as soluble folded molecules in *E. coli*.

Recently, Lindwall et al. (2000) undertook the first systematic investigation of the effect of solvents used in cell washing and lysis on the recovery of soluble proteins. They observed that the washing procedure for *E. coli* cells, as well as the lysis buffer used, affects the protein distribution. Bacterial growth medium or the intracellular milieu may contain molecules, large or small, that can affect the solubilization of proteins. This is one of the complicating factors in the isolation of KGFR-fusion proteins, which were found in large soluble complexes with

DNA. Washing cells containing overexpressed protein in a buffer of moderate ionic strength (e.g., 10 mM Tris, pH 8.5, 100 mM NaCl, 1 mM EDTA) before they are disrupted can help to decrease the concentration of media components and extracellular metabolic products. Employing a variety of additives, as described above, can also help break up interactions between overexpressed protein and extracellular components. The wash step was found to be essential for obtaining reproducible recovery of soluble expressed proteins. The cell-washing step is probably not important when inclusion bodies are the protein source. When the lysed cells are centrifuged to obtain the inclusion body pellet, any cell culture medium contamination remaining after the cells are collected is removed.

Lindwall et al. (2000) developed a so-called "sparse matrix" of buffer conditions for cell washing and lysis, grouping various solvent components into three classes, as shown in Table 1. A series of buffer solutions containing one choice from each list are generated and tested. The first is a list of buffers for varying the pH over the range where proteins are generally folded. The second is a list of salts appropriate for achieving a full range of chaotrope behavior at a variety of salt concentrations. The third is a list of additives that sometimes improve protein solubility in crude extracts. With this sparse matrix approach, together with cell washing, they were able to improve the soluble expression of the tubulin C-terminal fragment and the two forms of the avian sarcoma virus reverse transcriptase.

One may screen varying concentrations of non-chaotropic, slightly chaotropic and moderately chaotropic salts, as well as low concentrations of chaotropic salts, to find the best conditions. In a crude extract, the solubility

Table 1.
The Three Lists Used to Generate a Combination of Different Lysis Buffers[a]

List 1: Buffers	List 2: Stabilizing Solutes	List 3: Solubilizers
Tris, pH 8.2	L-glutamate	glycerol
Tris, pH 7.6	LiCl	isopropanol
HEPES, pH 7.0	$(NH_4)_2SO_4$	Triton X-100
triethanolamine, pH 8.5	KCl	CHAPS
K_2HPO_4/KH_2PO_4, pH 6.0	NaCl	n-octyl-β-D-glucoside
KH_2PO_4/K_2HPO_4, pH 4.3	K_2HPO_4/KH_2PO_4	deoxycholate
sodium acetate, pH 5.5	L-glutamine	dithiothreitol
		EDTA
		urea
		dextran sulfate
		$ZnCL_2$
		calcium acetate
		$MgSO_4$

[a]Modified from Lindwall et al. (2000)

optimum can perhaps best be thought of as a balance between protein aggregation and unfolding.

Other compounds can also have a positive effect on protein solubility. Some small organic molecules such as isopropanol, which might be denaturing at high concentrations, can be used to break up nonspecific aggregation at low concentrations (Asakura et al., 1978). Glycerol can promote solubilization and stability at the same time and is generally not considered to be denaturing at any concentration (Arakawa and Timasheff, 1985; Gekko and Timasheff, 1981). Polyvalent ions like spermidine or dextran sulfate (Hedman and Gustafsson, 1984) may be able to inhibit nonspecific interactions of the expressed protein with polysaccharides, nucleic acids, or abundant highly charged host proteins such as histones. Polyethylene imine or other polycations bind nucleic acids. Sugars may be able to act in a similar way and have the additional advantage of tending to stabilize protein folding (Lee and Timasheff, 1982; Arakawa and Timasheff, 1982a).

Different salts have widely varying effects on protein stability and solubility (Arakawa and Timasheff, 1982b, 1984). Some have varying effects depending on the nature of the individual protein (Arakawa et al., 1990). Detergents can also be helpful in overcoming aggregation, perhaps by binding to hydrophobic surface patches on the surface of the protein of interest (Bam et al., 1996; Charman et al., 1993). Detergents with high critical micellar concentrations can be particularly helpful since they can be removed later by dialysis, if desired.

PURIFICATION OF SOLUBLE, FOLDED PROTEINS

There are a number of procedures that have been used to purify soluble, folded proteins from *E. coli*, as summarized by Le and Trotta (1991). Crude cell lysate or *E. coli* cell culture media is often difficult to centrifuge or filter to obtain a clear supernatant. Therefore, the first chromatographic step is best carried out in a batch-mode. In this procedure, the lysate supernatant is mixed with a resin (ion-exchange chromatography, hydrophobic interaction chromatography, etc.) in an open container. The mixture is allowed to incubate for about 30 minutes, to allow time for the protein to bind to the resin. The entire suspension is then transferred to a filter, and extensively washed with the lysis or column buffer. After washing, the resin is transferred to an appropriate column for elution of the protein. In addition to being able to process a large volume of supernatant and wash the resin, to which the protein is bound, more extensively, this batch-mode binding has another advantage. When a sample is loaded onto a column by pumping, all the components can be bound to the top layer of the column resin. This process by itself often causes irreversible association and precipitation of the protein of interest due to the high protein concentration in the binding zone.

The same type of problem also occurs upon elution of the protein from the resin, particularly when the binding is tight, requiring high salt concentrations for elution. It should be noted that NaCl, the most commonly used salt for elution, is also a salting-out salt (Arakawa and Timasheff, 1984). High salt and high protein concentrations that might occur upon elution can cause aggregation and precipitation of the protein in the column. Batch-mode binding can avoid having a zone of high protein concentration. Whenever column-mode binding is used, it is always recommended that the column be washed at the end of elution with a salt/urea mixture, which should remove irreversibly aggregated materials trapped in the column.

Since the protein isolated from the soluble fraction is folded, any harsh, potentially denaturing conditions such as pH above pH 9 or below pH 5, or denaturant concentration above 2 M urea or 1 M guanidine HCL (GdnHCl), should be avoided during the purification process. However, if the folded protein is stable between pH 4–5, the solution can be titrated to this pH range, and a majority of the *E. coli* proteins will precipitate. Except where proteins are stable or reversibly denatured in extreme acid and strongly hydrophobic environments, reverse-phase chromatography purification should not be used unless absolutely essential.

PURIFICATION AND REFOLDING OF SOLUBLE, MISFOLDED PROTEINS

Occasionally, a protein located in the soluble fraction of the cell will still be misfolded. An example of this is agouti-related protein, which is expressed as a secreted protein *in vivo* (Rosenfeld et al., 1998). When a truncated mutant of this protein was expressed in *E. coli*, it was found in the soluble fraction. However, attempts to purify this protein demonstrated that it existed as large, disulfide-bonded aggregates. The cells were lysed in water, and the aggregates were solubilized by incubation in 6 M GdnHCl at room temperature. The protein was then refolded by dilution into an oxidation/refolding buffer containing 1 M urea, and a cystein/cystamine redox pair at pH 7.0. The refolded protein was then purified by conventional chromatography to yield highly pure, active protein. Another example of a misfolded, soluble protein is the KGFR-fusion protein mentioned above (M. Snavely, private communication). KGFR was expressed as a fusion protein in order to localize it in the soluble fraction, and indeed the protein was found in the cytosol. However, it existed as very large, soluble complexes containing DNA and covalently bound oligomers. Isolation of homogeneous protein from this source was not practical.

PURIFICATION AND REFOLDING OF PROTEINS FROM INCLUSION BODIES

Washing and Solubilization of Inclusion Bodies

Contaminating proteins and other materials can be removed by washing inclusion bodies with various aqueous solutions, containing mild detergents (at μM to mM concentrations), such as Triton X-100, Tween 20, Tween 80 or deoxycholate, and low concentrations of urea or GdnHCl, or EDTA. This washing step may be repeated, as long as the majority of the expressed protein stays in the insoluble fraction (see Table I of Fischer et al., 1993).

Inclusion bodies can then be solubilized with urea, GdnHCl or strong detergents, such as Sarkosyl or SDS (see Table II and III of Fischer et al., 1993), or combinations of these denaturants with extremes of pH (Krensky and Nelson, 1997; Khan et al., 1998). For example, inclusion bodies containing an EF-2 fragment were washed once with 1% Triton X-100, then with 0.25% Tween 20/1 mM EDTA, then with 10% DOC, and finally with distilled water (Ras and Bodley, 1996). The washed inclusion bodies were solubilized by the slow addition of 6 M urea, 50 mM Tris pH 8.0 and 5 mM β-mercatoethanol. Similarly, the link module protein inclusion bodies were solubilized in 8 M GdnHCl, 50 mM Tris pH 8.0, 30 mM dithiothreitol (DTT) and subjected to Sephacryl S100 gel filtration in 6 M GdnHCl and 20 mM Tris (pH 8.0), resulting in formation of two disulfide bonds (Day et al., 1996). Sigma transcription factor (40 K) was solubilized in 0.25% Sarkosyl and refolded by dialysis versus 50 mM Tris, pH 7.9, 10% glycerol, 0.1 mM EDTA and 0.1 mM DTT (Nguyen et al., 1996).

Urea or guanidine-HCl at high concentrations and strong detergents can solubilize most inclusion bodies and prevent re-precipitation of the solubilized proteins. Although producting visibly clear solutions, it is possible that the solubilized proteins exist as associated oligomers, even under these highly denaturing conditions. Such aggregation, if present, will usually cause problems during refolding.

Purification of Expressed Proteins from Inclusion Bodies

When urea is used for solubilization, any column chromatography, except hydrophobic interaction, can be used as the next step in the purification procedure. When ionic detergents or GdnHCl are used to dissolve inclusion bodies, size-exclusion chromatography (SEC) or Ni-affinity chromatography is appropriate. For disulfide-containing proteins, the inclusion of reducing agents may be necessary when performing column chromatography at alkaline pH, to keep the protein reduced. It is simpler if solubilization is performed at low pH, as no reducing agent is required under acidic conditions. Here again, the batch-mode procedure for binding the protein to the resin is highly recommended. For proteins with a pI above 7.5, the following general protocol can be used, which is diagrammed in Figure 2. Inclusion bodies are solubilized in 7 M urea, pH 3 to 6, by rigorous homogenization. After centrifugation the supernatant is batch-bound to a cation-exchange resin, and the resin is extensively washed on the filter with 7 M urea buffer at pH 3 to 6. Then, the bound proteins are eluted with NaCl in 7 M urea at pH 3 to 6. Since the pH is maintained below 6.0, no reducing agent is included. For acidic proteins, solubilization may be done at high pH, in the presence of a low concentration of reducing agent.

These partially purified materials usually require a refolding procedure. However, it is possible that none of these initial chromatography steps will lead to an increase in purity, but will result in material that can be used successfully in a refolding step. Murine IFN-γ (mIFN-γ) is an example of such a protein (unpublished observations). Application of the general protocol (Figure 2) to inclusion bodies of mIFN-γ, which had been solubilized in 7 M urea, resulted in no binding of mIFN-γ to either cation (low pH) or anion (high pH) exchange chromatography columns. In addition, attempts to directly refold the urea-

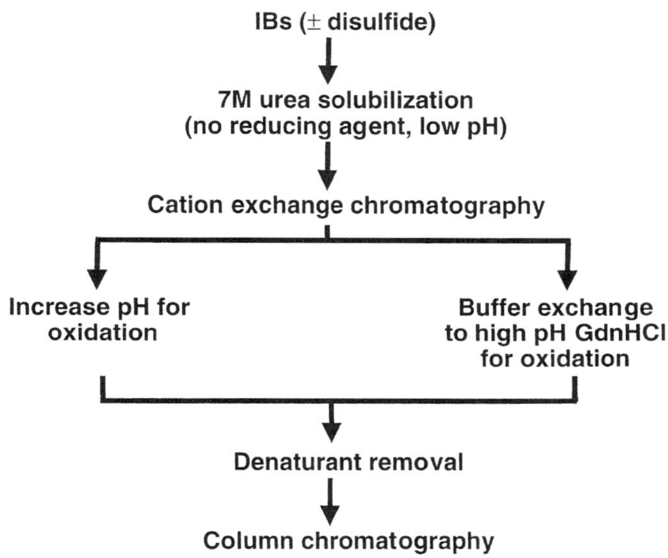

Figure 2. Generic purification protocol for recombinant proteins expressed in *E. coli* at inclusion bodies (IB's).

solubilized material were also unsuccessful. Rather, only solubilization by 6–7 M GdnHCl at pH 3–4.5 followed by SEC, in 5–6 M GdnHCl at pH 3–4.5, resulted in a partially purified material that was refolded successfully. It appears that mIFN-γ purification and refolding is hindered by a contaminant that can only be removed in the presence of GdnHCl. It is also possible that the urea-solubilized material, though forming a visibly clear solution, may still have been severely aggregated. In contrast, it appears that aggregation did not occur in the presence of GdnHCl.

REFOLDING MECHANISM

Protein folding in *E. coli* is mediated by molecular chaperones (Gething and Sambrook, 1992; Martin et al., 1991), and foldases, such as peptidyl-prolyl isomerase and protein disulfide isomerase, which catalyze proline isomerization and disulfide bond formation, respectively (Thomas et al., 1997; Ranson et al., 1998; Guise et al., 1996; Hendrick and Hartl, 1993). Therefore, *in vivo* protein folding has these two important elements: 1) aggregation of folding intermediates are minimized *in vivo* by molecular chaperones; and 2) disulfide formation is facilitated by protein disulfide isomerase. Inclusion bodies solubilized by detergents or denaturants, before and after partial purification, must be refolded by removing these solubilizing and denaturing reagents and inducing disulfide bonds, without the aid of these protein helpers. Due to the strong reducing environment (Freedman, 1992), disulfide bonds are almost never formed in inclusion bodies produced in *E. coli* (Tsuji et al., 1987). However, there are a few exceptions (Rosenfeld et al., 1998; Fromage et al., 1991; Sarmientos et al., 1989).

For proteins with no disulfide bonds in the native structure, protein refolding occurs by rapid formation of the native structure, either by direct conversion to the native conformation or via an intermediate conformation, as shown schematically shown in Figure 3. Refolding is induced by the removal of the denaturing agents that were used to solubilize inclusion bodies into the denatured "D" state. There are many ways to remove or reduce the denaturant concentration and these are described below. Since the conversion of D or the intermediate, I, to aggregates is irreversible, the manner in which the denaturant concentration is reduced has great impact on the rate of aggregation and, hence, the refolding efficiency (Mitraki et al., 1987; Brems, 1988; Mitraki et al., 1987; Brems, 1988; Mitraki and King, 1989; Oberg et al., 1994; Speed et al., 1996). Slow removal of denaturants or an equivalent procedure that increases the population of I for a prolonged period might allow I to proceed to the native state, N. The denaturant concentration must be such that I can convert to N. However, under these conditions a concentration of denaturant that favors a shift of the

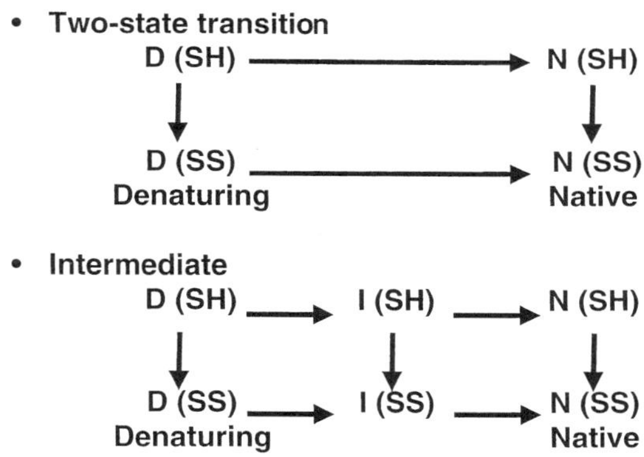

Figure 3. Refolding mechanism for proteins solubilized from inclusion bodies.

molecular population towards I might also promote formation of aggregates from I. Rapid dilution of the denaturant and the protein may avoid aggregation, if the I to N conversion can occur rapidly enough. If the I to N reaction is too slow, then aggregation predominates. The solubility and stability of I and the relative rates of the I $\rightarrow$ N reaction versus the I $\rightarrow$ aggregate reaction will determine the procedure by which the denaturant concentration should be reduced. Because formation of aggregates is a second- or higher-order process and refolding is a first order process, reducing protein concentration will tend to reduce aggregate formation. For disulfide-containing proteins, the refolding process includes the additional consideration of the formation of the correct disulfide bonds. As described in the next section, oxidation or disulfide exchange is used to modulate the rate of disulfide formation.

Although I (SH) shown in Figure 3 can be generated similarly to that of non-disulfide proteins, the I(SH) to I(SS) conversion and consequent conformational change are essential for formation of N in disulfide-containing proteins. Disulfide bond formation can be accelerated by disulfide exchange reagents. An optimal concentration, achieved by either dialysis or dilution, of denaturant may be that at which the solubility of I(SH) is maintained, yet at which I(SH) can assume a native-like structure to allow formation of the correct disulfide bonds.

As noted in Figure 3, I(SH) to I(SS) will be faster at lower denaturant concentrations. However, the solubility of I(SH) is lower at lower denaturant concentrations, enhancing aggregation. It is also possible that the conformation of I(SH) and I(SS) may be dependent on the denaturant concentration. In other words, there may be an optimal concentration of denaturant that increases the solubility of I(SH), thereby minimizing aggregation and facilitating the conversion of I(SH) to I(SS).

Detailed studies on reduced murine IL-6 demonstrated that soluble aggregates are generated upon removal of denaturant. These aggregates have enhanced fluorescence intensity, about 70% of the α-helical content of the native protein, and increased ANS binding, indicating loss of tertiary structure (Zhang et al., 1997). Despite the tendency to aggregate and possess minimal tertiary structure, upon dilution into an assay media murine IL-6 adopts the native disulfide structure and the fully active conformation. These results indicate that aggregation is at least partially reversible and that upon disaggregation this protein assumes an active conformation (Zhang et al., 1992).

Ovalbumin exhibits a different behavior. The native conformation has one disulfide and four sulfhydryl groups. When the fully reduced, the native or the non-native protein containing a single disulfide is denatured in 9 M urea and then diluted 20-fold into a neutral refolding buffer (a final urea concentration of 0.45 M), there is an initial burst of formation of an intermediate conformation with about 60% of the CD signal of the native protein, during which time disulfide rearrangement and oxidation take place (Onda et al., 1997; 1997; Tatsumi et al., 1994). This result implies that the folding intermediate can adopt different conformations, allowing it to rearrange the disulfide pairing, instead of forming soluble aggregates as was the case for murine IL-6. However, the single disulfide appears non-essential for refolding, since the reduced and single-disulfide native ovalbumin are similar in structure (Takahashi, 1991, 1992).

Studies of the unfolding and refolding of bovine pancreatic trypsin inhibitor (BPTI), a protein which contains three disulfide bonds in the native state, set the foundation for the analysis and understanding of the folding mechanism of disulfide-containing proteins (Creighton, 1977, 1978; Creighton and Goldenberg, 1984; Weissman and Kim, 1991; Zhang and Goldenberg, 1993; Ma and Anderson, 1997). With this protein, the fully reduced species and various folding intermediates are soluble and capable of disulfide exchange, ultimately resulting in the formation of the most stable three-disulfide species in aqueous buffer. A similar observation was made with insulin-like growth factor, which also contains three disulfide bonds (Rosenfeld et al., 1997). In this case, the folding intermediates appear to compete with the aggregation reaction.

The landmark studies on the protein folding pathways for the phage P22 tailspike protein also provide critical insight into protein refolding (Betts and King, 1988; Danner and Seckler, 1993; Speed et al., 1995; King et al., 1996; Mitraki et al., 1991). In this case, refolding of the protein denatured in 5 M urea was initiated by dilution to a final urea concentration of 0.8 M. Off-pathway aggregation was greatly reduced by incubating the refolded material at 0°C prior to raising the temperature to 20°C. This result suggests that the intermediate formed upon dilution further refolds at 0°C to a more stable structure (Betts and King, 1998).

Disulfide Bond Formation

There are a number of review articles that describe methods of facilitating or aiding in disulfide formation (e.g., Fischer et al., 1993; Mukhopadhyay, 1997); these methods are summarized in Table II. The simplest procedure is air oxidation, which is slow, but can be accelerated by the inclusion of Cu^{2+} ions in the micromolar range. Air oxidation can also be carried out in the presence of a free thiol, such as β-mercaptoethanol (Epstein and Goldberg, 1963).

There can be advantages in generating mixed disulfides before purification. Proteins in which all the sulfhydryl groups are reversibly modified are stable and have altered solubility. Although the structure of the modified protein, I(SSR), may be similar to I(SH), the former may be more soluble, which may allow for the use of lower denaturant concentrations. Reducing denaturant concentration can help to avoid disulfide exchange.

In an alternative approach, often a sufficient amount of reducing agent, such as reduced glutathione, in combination with its oxidized form, is used to initiate disulfide shuffling under refolding conditions. This method converts I(SH) to I(SS) and minimizes incorrect disulfide bond formation. Ejima et al. (1999) have shown that lower concentrations of reduced and oxidized glutathione generate more of the active disulfide structure of human IL-6. For all of these approaches, the concentration of denaturant plays an essential role, because it determines the conformation, solubility and flexibility of the intermediate structures in which disulfide bonds are formed.

Removal of Denaturant

For disulfide-containing proteins, oxidation may be performed before or during refolding. However, for some proteins after the correct disulfides have been formed, the manner in which denaturants are removed can affect the final refolded structure. The folding of non-disulfide proteins is determined almost certainly by the process of denaturant removal. Renaturation occurs when solubilizing agents (denaturants) are removed by dilution, dialysis, diafiltration or solid phase exchange.

Dialysis. The rate of the removal of denaturants is determined by the mode of dialysis or diafiltration. If dialysis is done against a step-wise descending denaturation concentration (e.g. 7 to 5 to 3 M urea as shown in Fig. 4), then an equilibrium in solute concentration is reached at each denaturant concentration, which can allow for equilibrium distribution of protein molecules in different states. However, there can be kinetic competition between aggregation and

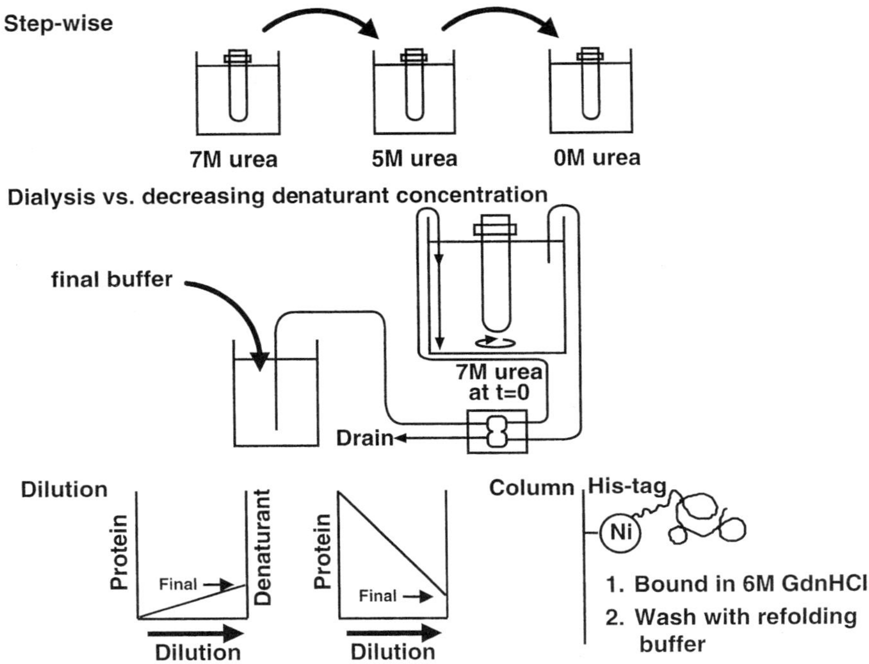

Figure 4. Diagrams of different protocols to reduce denaturant concentration.

refolding, such that the equilibrium between protein states is not achieved. One-step dialysis against buffer can also result in solution conditions in which this kinetic competition can occur.

Another method is to perform dialysis versus descending denaturation concentration with time. This type of dialysis against a concentration gradient can be accomplished using a gradient former or a pump shown in Figure 4. Dialysis tubing containing solubilized proteins is placed in the initial solvent (e.g. 7 M urea), and the final refolding buffer is gradually pumped into the initial solvent. The rate of change in denaturant concentration can be controlled by adjusting the flow rate of the pump. This dialysis method was successfully used for redox-coupled refolding of lysozyme (Maeda et al., 1995), immunoglobulin (Maeda et al., 1996) and platelet-derived growth factor-BB (PDGF-BB) (Alexander et al., 1992).

Dilution. Dilution is carried out by either diluting the protein solution in denaturant into a refolding solvent containing lower or no denaturant, or by adding refolding solvent into the protein solution. In the first case, the protein concentration starts at zero and gradually increases to the final concentration while the denaturant concentration is initially that of the refolding solvent and gradually increases. If the refolded structure is oligomeric, then this method of dilution may not work well, because the low protein concentration present at the onset of dilution may be too low to allow formation of stable oligomeric structures. In the second case, both protein and denaturant concentration decrease as the solvent is added. The choice of dilution method will depend on the folding kinetics and aggregation tendency of the intermediate species.

In general, dilution works well for disulfide-free proteins or disulfide-containing proteins whose disulfides are non-essential for protein structure and refolding (Arakawa et al., 1985; Takahashi and Hirose, 1992; Takahashi et al., 1991). As described above, this type of protein can acquire the native conformation spontaneously upon denaturant removal, and slow removal may allow sufficient time for aggregation of the intermediate structures.

A large number of non-disulfide proteins including FGFs, IFN-γ and rat intestinal fatty acid binding protein have been refolded by dilution. The last protein, an all-β-type protein, undergoes a transition from unfolded protein through intermediate structures to native protein upon dilution into refolding buffer, with higher protein concentration ($>20\,\mu M$) resulting in aggregation and precipitation (Ropson et al., 1990).

Solid Phase Removal of Denaturants. Although seldom used for commercial production of recombinant proteins, renaturation can be performed by exchanging the solvent while a protein is bound to a column. A protein is bound to a column in the presence of the denaturant that is used to solubilize the protein. Histidine fusion tags are often used, because the tagged protein can bind to a nickel resin in concentrated denaturant. The use of the tag for binding rather than a region of the mature protein itself allows the protein to have a flexible structure for refolding. After the protein is bound to the resin, the denaturant is exchanged with refolding buffer. This procedure has also been applied successfully to refolding of α-glucosidase on an ion exchange column, using the addition of a hexa-arginine tag to immobilize the protein (Stampfer et al., 1996).

Gel filtration. Refolding can be initiated by injecting a denatured protein in urea or GdnHCl onto a gel filtration column equilibrated with refolding buffer (Werner et al., 1994; Batas and Chaudhuri, 1999; Khan et al., 1998). It is essential that not only the elution buffer is optimal for refolding but also that it is compatible with the elution of refolded proteins. Dilution of both protein and denaturant concentration occurs simultaneously. Also, refolded proteins are somewhat dispersed in the gel matrix, which aids in preventing aggregation and helps separate refolded proteins from misfolded aggregates. A successful refolding of C1-inhibitor from the SDS-denatured state was achieved by this method (Gauthier and Patston, 1997).

Effects of Tag Sequences

To facilitate purification, as well as expression and solubility of proteins, a tag-sequence such as hexa-His, hexa-Arg, Flag-tag, glutathione S transferase (GST) or maltose binding protein (MBP) can be incorporated at the N- or C-terminal end of the mature protein. As expected, linking soluble cytoplasmic proteins with the tag sequence usually results in soluble fusion proteins. It should be noted that these tag sequences are often inaccessible to the affinity resin and, therefore, ineffective for purification.

MBP is also used to direct MBP proteins into the periplasmic space through a specific signal sequence for periplasmic transport (see above). For example, hybrid expression of MBP with CD4 fragments also resulted in periplasmic expression of the CD4 proteins (Clement et al., 1996). MBP fusion has been successfully used for generating the fragments of the G-CSF receptor (unpublished observations). Purification and protease cleavage resulted in various forms of the receptor, which were competent for binding its ligand, G-CSF. MBP was also fused to C-terminus of pepsinogen, which resulted in inclusion bodies. However, MBP fusion facilitated refolding of pepsinogen for an unknown reason (Sachdev and Chirgwin, 1998).

GST fusion proteins form a dimer through GST. GST is unstable upon dissociation. Hence, if the fusion protein interferes with the dimerization of GST, then the GST portion may become unstable, causing the fusion protein to precipitate. Also, because of the inability of fused proteins to fold into stable structures in *E. coli* cytoplasm, fusion proteins often appear as insoluble inclusion bodies, as is the case for GST-PrPc fusion (Völkel et al., 1998).

Comparison of three fusion partners, MBP, GST and thioredoxin, using widely different proteins expressed in *E. coli*, suggested that MBP is most effective in solubilizing proteins that are "insoluble" when expressed without fusion partners (Kapust and Waugh, 1999). Refolding of GST fusion proteins made with extracellular proteins has not yet been successful, although dimeric GST itself can be refolded from precipitates.

Effects of Excipients

Refolding of proteins is initiated by decreasing the denaturant concentrations. There have been some attempts to increase the refolding efficiency using folding enhancers including sugars, amino acids, surfactants and polymers (Cleland, 1993). ArgHCl has been used as an excipient to facilitate the refolding process (Buchner and Rudolph, 1991; Rudolph et al., 1992; Rinas et al., 1990; Ahn et al., 1997; Arora and Khanna, 1996; Buchner et al., 1992), although there are only a few studies that compare refolding yield in the absence and presence of this reagent. A 10-fold increase in the yield of IFN-γ was observed with 0.5 M L-arginine, compared to renaturation in its absence (Arora and Khanna, 1996). Rudolph et al. (1992) argue that L-arginine has a chaotropic nature and, hence, destabilizes incorrectly folded structures, allowing the molecules to proceed down the pathway to correct folding. However, little is known about the interactions of this amino acid with proteins. The observation that ArgHCl can substitute at least partly for the effect of GdnHCl suggests that it acts on proteins by a mechanism similar to that of GdnHCl or urea. These compounds have two effects, hydrogen bonding to peptide groups and solubilization of non-polar compounds. No data are available on the effects of ArgHCl on the capacity to form hydrogen bonds to peptide groups or on the solubility of non-polar compounds, such as the solubility measurements of amino acids and transfer free energy of side chains, except for the preferential interaction measurements of ArgHCl with bovine serum albumin (BSA) and lysozyme (Kita et al., 1994).

Figure 5 shows the plot of the preferential interaction parameter for ArgHCl with BSA, $\left(\frac{\partial g_1}{\partial g_2}\right)$, as a function of ArgHCl concentration. This value is initially negative at low concentrations, crosses over zero and becomes positive at higher concentrations. Let us examine the meaning of $\left(\frac{\partial g_1}{\partial g_2}\right)$ in more detail. This parameter is related to absolute hydration (A_1) and solute binding (A_3) of the

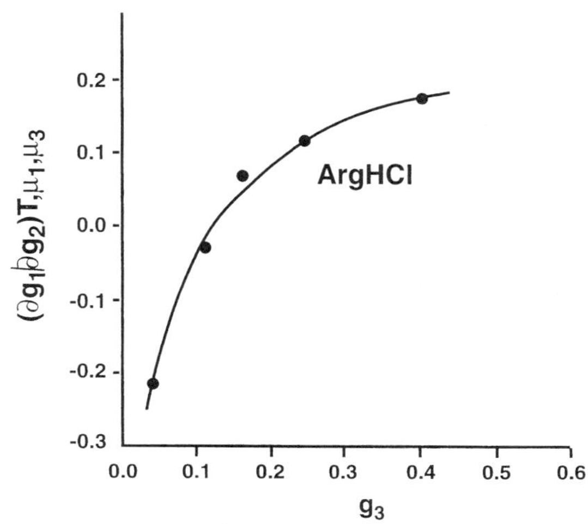

Figure 5. Preferential hydration of bovine serum albumin in argininie-HCl. Data taken from Arakawa and Timasheff, 1982b.

protein by the equation $\left(\frac{\partial g_1}{\partial g_2}\right) = A_1 - \frac{1}{g_3} A_3$, where g_3 is the concentration of solute. Assuming A_1 is constant as ~0.3 g water per g protein, the negative value of $\left(\frac{\partial g_1}{\partial g_2}\right)$ means that A_3 is positive and the term A_3/g_3 is greater than A_1 at low concentrations, as shown in Figure 6. Since BSA is in the native state in the presence of ArgHCl, this result indicates that ArgHCl can bind to the native state of the protein. As the ArgHCl concentration increases, $\left(\frac{\partial g_1}{\partial g_2}\right)$ becomes positive, which indicates that A_3/g_3 becomes vanishingly small. Taking all these results together, ArgHCl appears to have a weak, but saturating binding to BSA. Therefore, it may be possible for ArgHCl to bind to the native-like structures found during the refolding process and stabilize them.

Polyethylene glycol (PEG) has been used to inhibit aggregation of folding intermediates that appear during the refolding process of carbonic anhydrase and bovine growth hormone (Cleland and Wang, 1990; Cleland et al., 1992a; 1992b; Cleland and Randolph, 1992; Cleland, 1993). PEG has two opposing effects, a salting-out effect due to excluded volume and hydrophobic binding to proteins due to its surfactant character (Arakawa and Timasheff, 1985; Kita et al., 1994; Lee and Lee, 1979, 1981). The concentration of PEG used in the refolding studies was much lower than that required for salting-out, and hence the mechanism of

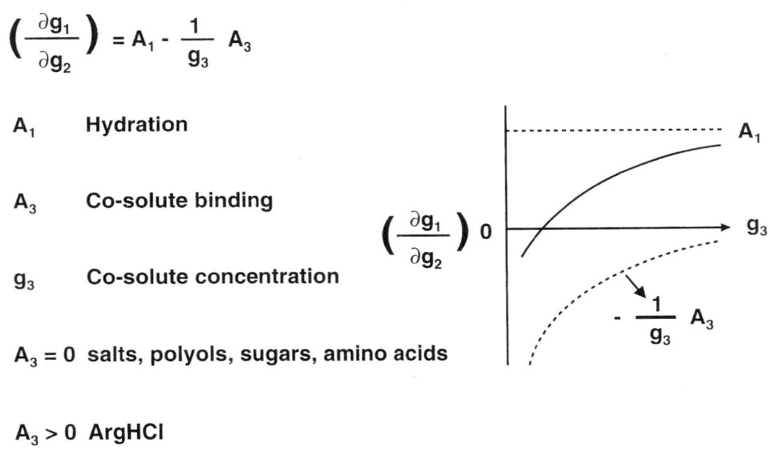

Figure 6. Schematic illustration of hydration and arginine-HCl binding to BSA used to interpret the observed preferential hydration of BSA in arginine-HCl shown in Figure 5.

inhibition of protein aggregation is due to hydrophobic interactions between PEG and partially-folded protein species.

Other additives such as sugars, salts, polyols, amines, alcohols and detergents have also been used to inhibit protein aggregation during refolding. In general, stabilizing salts, polyols, sugars, and amines (e.g., trimethylamine-N oxide) increase the stability of more compact structures. Recently some of these compounds were also shown to counteract the destabilizing action of urea (Bolen's and Lin's). This effect may be used to induce folding, rather than decreasing the denaturant concentration to initiate refolding. It is also becoming apparent that some of these stabilizing solutes can increase the average compactness of protein molecules (e.g., Kendrick et al., 1998). This effect is most likely due to the presence of multiple states of native-like structure with varying degrees of molecular compactness in the refolding mixture. These excipients could be affecting the distribution of these multiple states to favor the more compact states. Detergents have also been used to assist in refolding, presumably by transiently interacting with partially folded protein molecules and sterically hindering the intermolecular contact required for aggregation (Zardeneta and Horowitz, 1994a; 1994b; Horowitz, 1993).

French and Schmid (1995) have shown that the addition of concentrated NaCl can induce native-like structure in a single disulfide mutant of RNase T_1 in aqueous solution. Formation of the remaining disulfide is then strongly enhanced relative to its formation in the unfolded structure in the absence of NaCl. Reduced and carboxymethylated ribonuclease T_1 also showed refolding into a native-like structure in the presence of 2.7 M TMAO, as determined by gel filtration chromatography (protein concentration at 20 μg/ml), fluorescence spectroscopy (0.3 mg/ml) and CD spectroscopy (1.5 mg/ml) (Baskakov and Bolen, 1998). At 0.1–0.2 mg/ml, reduced and carboxymethylated ribonuclease A also exhibited substantial compaction, approaching the size of the native protein in the presence of 1 M sucrose, sarcosine or TMAO (Qu et al., 1998). It would be interesting to see if a fully-reduced protein with a similar folded structure can efficiently lead to the native disulfide structure. However, it should be emphasized that most proteins, in the absence of denaturant and disulfide bonds, have some tertiary structure.

Response Surface Methodology

Response surface methodology (RSM, an experimental design technique) was used to investigate optimal refolding conditions of *Pseudomonas fluorescens* lipase, which was expressed in the inclusion bodies of *E. coli* (Ahn et al., 1997).

Three interacting factors—protein concentration, pH and GdnHCl concentration—were selected as the variables of RSM. The protein concentration did not affect the refolding yield within the selected range (50–340 μg/ml) at low temperatures of 4°C and −15°C, but was a critical factor at 25°C where refolding yield significantly decreased with increasing protein concentration. GdnHCl concentration and pH were significant factors under all experimental conditions, although there was no obvious trend towards an optimal pH. Depending on temperature and additives, the optimum was simply at neutral pH. Glycerol shifted the optimal GdnHCl concentration to higher values, whereas arginine shifted it to the lower concentrations. In this study, RSM made it possible to investigate successfully the optimal conditions of *in vitro* refolding and to elucidate interactions between refolding factors, with a minimum number of experiments. Under the optimal conditions determined by RSM, an approximately 90% refolding yield was obtained, a 30% increase over that realized with the conventional method.

High Pressure Disaggregation and Refolding

Recently, an alternative disaggregation and refolding procedure that employs high hydrostatic pressure has been described (St. John et al., 1999). Application of high pressure is used to solubilize proteins from aggregates or inclusion bodies. Under the appropriate pressure regime the molecules can then refold, without reaggregating. This method requires relatively low concentrations of denaturants or no denaturants to solubilize proteins; the denaturants, if used, do not need to be removed prior to refolding. Another major advantage of high pressure processing over conventional refolding methods is that recovery of native protein appears to be insensitive to protein concentration. Yields of native protein of 80–100% have been achieved with protein concentrations of 10 mg/ml or greater. With a redox mixture, this approach has even been applied to aggregates that are covalently linked by disulfide bonds.

METHODS TO ANALYZE FOLDED STRUCTURES

The above sections have discussed ways of optimizing the recovery of folded protein expressed in *E. coli*. Initially, improvements in the recovery of folded protein can be inferred from improvements in the solubility of the protein. However, eventually a more rigorous characterization of protein structure and function must be performed to determine how closely the refolded protein resembles that of the natural, native protein. Some techniques that are frequently employed for this comparison include bioactivity measurement, determination of disulfide bond structure, CD, FTIR, and fluorescence spectroscopies, SDS-PAGE and other electrophoretic techniques, limited proteolysis, and chromatographic analysis. A brief review of the application of these techniques to the analysis of proteins during and after folding is given below. For all of these techniques, the comparison of the *E. coli*-derived, folded protein to that of the natural, native protein is the most definitive in demonstrating that the recombinant protein has the appropriate structure and function. If the natural protein is not available, comparison to the protein derived from mammalian cells, where refolding was not involved, can also be used. When the mammalian counterpart is not available and when expressed proteins distribute into both soluble and insoluble fractions, the soluble protein may be compared to the refolded material to assess refolding.

Bioactivity

When the activity of a protein is known, and there is a bioassay readily available, the comparison of the activity of the *E. coli*-derived proteins to that of the natural protein, or at least the protein expressed in mammalian cells, provides a good indication of how successful the refolding procedure has been. The recovery of biological activity is the primary assay used to follow protein folding in many of the papers discussed in the section above. However, when naturally-derived proteins are glycosylated, and glycosylation is essential for bioactivity, a bioassay may be misleading, since even correctly folded *E. coli* proteins may be inactive due to lack of glycosylation (e.g., erythropoietin) (Goldwasser et al., 1974; Takeuchi et al., 1989; Goto et al., 1988).

Binding to Receptors

Surface plasmon resonance was used in studies on Human Fc, where decreased binding to the human Fc receptor was the first indication that the protein purified from the periplasm of two different *E. coli* strains had differences in their final structure relative to that of the natural, native protein (Wypych, unpublished results). Flow cytometry has also been employed to assess binding of refolded proteins, in this case to authentic cell-surface receptors. Kurucz and colleagues (1995) used this technique to develop a successful refolding scheme for Fv proteins.

Dilsulfide Bond Analysis

As described above, proteins derived from *E. coli* may require disulfide bond formation to have a stable folded structure. Therefore, determination of the disulfide bonding pattern in the final product is commonly employed as another demonstration of proper folding and oxidation. Disulfide bonds can be determined by peptide mapping or more readily by reverse-phase HPLC. In general, the elution position in reverse-phase HPLC shifts to a lower organic solvent concentration upon formation of correct disulfide bonds.

Non-reducing SDS-PAGE can also be used to follow disulfide bond formation. For large or average sized globular protein, formation of intra-molecular disulfides results in faster mobility, while small proteins often have reduced mobilities upon disulfide formation. The use of disulfide bond structural determination to follow refolding was demonstrated elegantly in the work of Creighton et al. (1977, 1978, 1989) and others (e.g., Weisman and Kim, 1991) on BPTI. This technique was also used in studies following the refolding and oxidation of IL-6 (Zhang et al., 1992), Fv (Kurucz et al., 1995) and ovalbumin (Takahashi 1991, 1992), and can be found throughout the literature. However, a correct disulfide structure does not necessarily correlate with correct refolding. Even with the correct disulfide bonding pattern the protein can sometimes misfold and form aggregate or have local regions of incorrect tertiary structure.

Spectroscopy

It is not always possible to measure the activity of a refolded protein, and often there is no natural reference protein available to compare to a protein isolated following cloning into *E. coli*. In this case, other tools must be employed to determine if the recombinant protein has a regular three-dimensional structure. Various spectroscopic techniques can be used for structural determination. CD, fluorescence and FTIR spectroscopies are the techniques most often employed to follow the course of protein refolding, and to assess the final structure. Far UV CD and FTIR spectroscopies can provide information on the secondary structure of proteins, whereas near UV CD and intrinsic fluorescence spectroscopies, and fluorescent dye binding assays can be used to provide information about the tertiary structure.

These techniques have commonly been used to follow protein folding pathways. For example, studies with IL-6 (Zhang et al., 1992) and ovalbumin (Onda et al., 1997; Tatsumi et al., 1994) employed CD spectroscopy to compare the structure of folding intermediates to the native structure of the protein. ANS binding, a technique that takes advantage of the increased fluorescence upon binding to hydrophobic patches on the surface of proteins (Arakawa et al., 1991), was also employed to follow changes in the structure of IL-6. CD, FTIR, and fluorescence spectroscopies were all employed to analyze the difference in structure between Fc cloned into different strains of *E. coli* (unpublished observations).

Conformational Stability

Spectroscopic signals often arise from association of non-native structures, rather than from folded structures. In this case, structural perturbation techniques can distinguish distinct folded structures from misfolded ones. The misfolded structure often gives rise to a monotonic change in signal as a function of strength of perturbation (e.g., temperature or concentration of denaturant), rather than the cooperative transition during unfolding of an initially native conformation. Heat, pH or denaturant-induced unfolding can be used to probe protein conformational stability. CD and other spectroscopies can be used to follow these changes in structure. Differential scanning calorimetry (DSC) can also be employed successfully. Snyder et al. (1996) used DSC to demonstrate that the refolded CMV proteinase had a different conformation than the molecule isolated as a soluble protein.

Limited Proteolysis

Limited proteolysis is another useful technique for probing the tertiary structure of a protein. This technique is not dependent on the presence of particular chromophores in the protein, but instead compares the susceptibility of proteins to proteolytic degradation (Fontana et al., 1997; Hubbard, 1998). This property is a reflection of the flexibility or stability of a protein. This technique can be used not only to assess degree of folding but also to determine the relative conformational mobility of a protein (e.g., Arakawa et al., 1990).

REFERENCES

Abrahmsen, L., Moks, T., Nilsson, B., and Uhlen, M., 1986. Secretion of heterologous gene products to the culture medium of Escherichia coli. *Nucleic Acids Res.* 14:7487.

Ahn, J.H., Lee, Y.P., and Rhee, J.S., 1997. Investigation of refolding condition for Pseudomonas fluorescens lipase by response surface methodology. *J. Biotech.* 54:151.

Alexander, D.M., Hesson, T., Mannarino, A., Cable, M., and Dalie, B.L., 1992. Isolation and purification of a biologically active human platelet-derived growth factor BB expressed in Escherichia coli. *Prot. Exp. Purif.* 3:204.

Arakawa, T., Kita, Y., and Narhi, L.O., 1991. Protein-ligand interaction as a method to study surface properties of proteins. *Meth. Biochem. Anal.* 35:87.

Arakawa, T., Visger, M.V., McGinley, M., Rohde, M.F., Fox, G.M., and Narhi, L.O., 1990. Alteration in folding efficiency and conformation of recombinant human tumor necrosis factor-alpha by replacing cysteines 69 and 101 with aspartic acid 69 and arginine 101. *Prot. Engineer.* 3:721.

Arakawa, T. and Timasheff, S.N., 1985. Theory of protein solubility. *Methods Enzymol.* 114:49.

Arakawa, T. and Timasheff, S.N., 1982a. Stabilization of protein structure by sugars. *Biochemistry* 21:6536.

Arakawa, T. and Timasheff, S.N., 1982b. Preferential interactions of proteins with salts in concentrated solutions. *Biochemistry* 21:6545.

Arakawa, T. and Timasheff, S.N., 1984. Mechanism of protein salting in and salting out by divalent cation salts: balance between hydration and salt binding. *Biochemistry* 23:5912.

Arakawa, T., Bhat, R., and Timasheff, S.N., 1990. Why preferential hydration does not always stabilize the native structure of globular proteins. *Biochemistry* 29: 1924.

Arakawa, T. and Timasheff, S.N., 1985. Mechanism of poly(ethylene glycol) interaction with proteins. *Biochemistry* 24:6756.

Arakawa, T., Alton, N.K., and Hsu, Y.-R., 1985. Preparation and characterization of recombinant DNA-derived human interferon-gamma. *J. Biol. Chem.* 260:14435.

Arakawa, T., Horan, T.P., Narhi, L.O., Rees, D.C., Schiffer, S.G., Holst, P.L., Prestrelski, S.J., Tsai, L.B., and Fox, G.M., 1993. Production and characterization of an analog of acidic fibroblast growth factor with enhanced stability and biological activity. *Prot. Engineer.* 6:541.

Arora, D. and Khanna, N., 1996. Method for increasing the yield of properly folded recombinant human gamma interferon from inclusion bodies. *J. Biotechnol.* 52:127.

Asakura, T., Adachi, K., and Schwartz, E., 1978. Stabilizing effect of various organic solvents on proteins. *J. Biol. Chem.* 253:6423.

Bam, N.B., Cleland, J.L., and Randolph, T.W., 1996. Molten globule intermediate of recombinant human growth hormone: stabilization with surfactants. *Biotech. Prog.* 12:801.

Baskakov, I. and Bolen, W., 1998. Forcing thermodynamically unfolded proteins to fold. *J. Biol. Chem.* 273:4831.

Batas, B. and Chaudhuri, J.B., 1999. Consideration of sample application and elution during size-exclusion chromatography based protein refolding. *J. Chromatogr.* A. 864:229.

Betts, S.D. and King, J., 1998. Cold rescue of the thermolabile tailspike intermediate at the junction between productive folding and off-pathway aggregation. *Prot. Sci.* 7:1516.

Brems, D.H., 1988. Solubility of different folding conformers of bovine growth hormone. *Biochemistry* 27:4541.

Buchner, J., Pastan, I., and Brinkmann, U., 1992. A method for increasing the yield of properly folded recombinant fusion proteins: Single-chain immunotoxins from renaturation of bacterial inclusion bodies. *Anal. Biochem.* 205:263.

Buchner, J. and Rudolph, R., 1991. Renaturation, purification and characterization of recombinant F(ab)-fragments produced in *E. coli. Bio/Technology* 9:157.

Caspers, P., Stieger, M., and Burn, P., 1994. Overproduction of bacterial chaperons improves the solubility of recombinant protein tyrosine kinase in *E. coli. Cell. Mol. Biol.* 40:635.

Chan, S.J., Weiss, J., Konrad, M., White, T., Bahl, C., Yu, S.-D., Mark, D., and Steiner, D.F., 1981. Biosynthesis and periplasmic segregation of human proinsulin in Escherichia coli. *Proc. Natl. Acad. Sci. USA* 78:5401.

Charman, S.A., Mason, K.L., and Charman, W.N., 1993. Techniques for assessing the effects of pharmaceutical excipients on the aggregation of porcine growth hormone. *Pharm. Res.* 10:954.

Chen, B.-L., Arakawa, T., Morris, C.F., Kenney, W.C., Wells, C.M., and Pitt, C.G., 1994. Aggregation pathway of recombinant human keratinocyte growth factor and its stabilization. *Pharm. Res.* 11:1581.

Chung, C.H., Ives, H.E., Almeda, S., and Goldberg, A.L., 1983. Purification from Escherichia coli of a periplasmic protein that is a potent inhibitor of pancreatic proteases. *J. Biol. Chem.* 258:11032.

Cleland, J.L., 1993. Impact of protein folding on biotechnology, in: *Protein Folding*, ACS Symposium Series Volume 526, J.L. Cleland, ed., ACS, Washington, D.C.

Cleland, J.L. and Wang, D.I.C., 1990. Cosolvent assisted protein refolding. *Bio/ Technology.* 8:1274.

Cleland, J.L., Hedgepeth, C., and Wang, D.I.C., 1992b. Polyethylene glycol enhanced refolding of bovine carbonic anhydrase B. Reaction stoichiometry and refolding model. *J. Biol. Chem.* 267:13327.

Cleland, J.L. and Randolph, T.W., 1992. Mechanism of polyethylene glycol interaction with the molten globule folding intermediate of bovine carbonic anhydrase B. *J. Biol. Chem.* 267:3147.

Cleland, J.L., Builder, S.E., Swartz, J.R., Winkler, M., Chang, J.Y., and Wang, D.I.C., 1992a. Polyethylene glycol enhanced protein refolding. *Bio/Technology* 10:86.

Clement, J.M., Jehanno, M., Popescu, O., Saurin, W., and Hofnung, M., 1996. Expression and biological activity of genetic fusions between MalE, the maltose binding protein from Escherichia coli and portions of CD4, the T-cell receptor of the AIDS virus. *Prot. Exp. Purif.* 8:319.

Collmer, A., Schoedel, C., Roder, D.L., Ried, J.L., and Rissler, J.F., 1985. Molecular cloning in Escherichia coli of Erwinia chrysanthemi genes encoding multiple forms of pectate lyase. *J. Bacteriol.* 161:913.

Creighton, T.E., 1977. Conformational restrictions on the pathway of folding and refolding of the pancreatic tyrosine inhibitor. *J. Mol. Biol.* 113:275.

Creighton, T.E. and Goldenberg, D.P., 1984. Kinetic role of a meta-stable native-like two-disulphide species in the folding transition of bovine pancreatic trypsin inhibitor. *J. Mol. Biol.* 179:497.

Danner, M. and Seckler, R., 1993. Mechanism of phage P22 tailspike protein folding mutations. *Protein Sci.* 2:1869.

Day, A.J., Aplin, R.T., and Willis, A.C., 1996. Overexpression, purification, and refolding of link module from human TSG-6 in Escherichia coli: effect of temperature, media, and mutagenesis on lysine misincorporation at arginine AGA codons. *Prot. Exp. Purif.* 8:1.

Denefle, P., Kovarik, S., Ciora, T., Gosselet, N., Benichou, J.-C., Latta, M., Guinet, F., Ryter, A., and Mayaux, J.-F., 1989. Heterologous protein export in *E. coli*: Influence of bacterial signal peptides on the export of human interleukin 1 beta. *Gene* 85:499.

Danilenko, D.M., Montestruque, S., Philo, J.S., Li, T., Hill, D., Speakman, J., Bahru, M., Zhang, M., Konishi, M., Itoh, N., Chirica, M., Delaney, M., Hernday, N., Martin, F., Hara, S., Talvenheimo, J., Narhi, L.O., and Arakawa, T., 1999. Recombinant rat fibroblast growth factor-16: structure and biological activity. *Arch. Biochem. Biophys.* 361:34.

Duffaud, G.D., March, P.E., and Inoyue, M., 1987. Expression and secretion of foreign proteins in *E. coli. Methods Enzymol.* 153:492.

Ejima, D., Watanabe, M., Soto, Y., Date, M., Yamada, N., and Takahara, Y., 1999. High yield refolding and purification process for recombinant human interleukin-6 expressed in Escherichia coli. *Biotech. Bioengineer.* 62:301.

Epstein, G.J. and Goldberg, R.A., 1963. A study of factors influencing the reactivation of reduced egg white lysozyme. *J. Biol. Chem.* 238:1380.

Fischer, B., Sumner, I., and Goodenough, P., 1993. Isolation, renaturation, and formation of disulfide bonds of eukaryotic proteins expressed in *E. coli* as inclusion bodies. *Biotech. Bioengineer.* 41:3.

Fletcher, R.S., Holleschak, G., Nagy, E., Arion, D., Borkow, G., Gu, B., Wainberg, M.A., and Parniak, M.A., 1996. Single-step purification of recombinant wild-type and mutant HIV-1 reverse transcriptase. *Prot. Exp. Purif.* 7:27.

Fontana, A., Polverino de Laureto, P., De Filippis, V., Saramella, E., and Zambonin, M., 1997. Probing the partly folded states of proteins by limited proteolysis. *Fold. Des.* 2:R17.

Fox, G.M., Schiffer, S.G., Rohde, M.F., Tsai, L.B., Bank, A.R., and Arakawa, T., 1988. Production, biological activity, and structure of recombinant basic fibroblast growth factor and an analog with cysteine replaced by serine. *J. Biol. Chem.* 263:18452.

Freedman, R.B., 1991. Protein folding in the cell, in: *Protein Folding* T.E. Creighton, ed., W.H. Freeman and Co., New York.

Frech, C. and Schmid, F.X., 1995. Influence of protein conformation on disulfide bond formation in the oxidative folding of ribonuclease T1. *J. Mol. Biol.* 251:135.

Fromage, N., Denefle, P., Cambou, B., Duchese, M., Joyeux, C., Kovarik, S., Marin, J., Imbault, F., Uzan, A., and Cartwright, T., 1991. Synthesis, purification and biological properties of a truncated mutant form of human tissue plasminogen activator produced in *E. coli. Fibrinolysis* 5:187.

Gauthier, M. and Patston, P.A., 1997. Reactivation of C1-inhibitor polymers by denaturation and gel-filtration chromatography. *Anal. Biochem.* 248:228.

Gekko, K. and Timasheff, S.N., 1981. Mechanism of protein stabilization by glycerol: Preferential hydration in glycerol-water mixtures. *Biochemistry* 20:4667.

Gething, M.J. and Sambrook, J., 1992. Protein folding in the cell. *Nature* 355:33.

Goto, M., Akai, K., Murakami, A., Hashimoto, C., Tsuda, E., Ueda, M., Kawanishi, G., Takahashi, N., Ishimoto, A., Chiba, H., and Sasaki, R., 1988. *Biotechnology* 6:67.

Ghrayeb, J., Kimura, H., Takahara, M., Hsiung, H., Masui, Y., and Inouye, M., 1984. Secretion cloning vectors in Escherichia coli. *EMBO J.* 3:2437.

Goldwasser, E., Kung, C.K.-H., and Eliason, J., 1974. On the mechanism erythropoietin induced differenciation B. The role of sialic acid in erythropoietin action. *J. Biol. Chem.* 249:4202.

Guise, A.D., West, S.M., and Chaudhuri, J.B., 1996. Protein folding in vivo and renaturation of recombinant proteins from inclusion bodies. *Mol. Biotech.* 6:53.

Halfmann, G., Brailly, H., Baradac, A., Montero-Julian, F.A., Lazdunski, C., and Baty, D., 1993. Targeting of interleukin-2 to the periplasm of Escherichia coli. *J. Gen. Microbiol.* 139:2465.

Hedman, P.O. and Gustafsson, J.G., 1984. Protein adsorbents intended for use in aqueous two-phase systems. *Anal. Biochem.* 138:411.

Hendrick, J.P. and Hartl, F.-U., 1993. Molecular chaperone functions of heat-shock proteins. *Ann. Rev. Biochem.* 62:349.

Hiraoka, O., Anaguchi, H., and Ota, Y., 1995. Formation of 1:1 complex of the cytokine receptor homologous region of granulocyte colony-stimulating factor receptor with ligand. *Biosci. Biotechnol. Biochem.* 59:2351.

Horuk, R., Reilly, D., and Yansura, D., 1997. Expression, purification, and characterization of Escherichia coli-derived recombinant human melanoma growth stimulating activity. *Meth. Enzymol.* 287:3.

Hu, M.C.T., Oiu, W.R., Wang, Y.-P., Hill, D., Ring, B.D., Scully, S., Bolon, B., DeRose, M., Leuthy, R., Simonet, W.S., Arakawa, T., and Danilenko, D.M., 1998. FGF-18, a novel member of the fibroblast growth factor family, stimulates hepatic and intestinal proliferation. *Mol. Cell Res.* 18:6063.

Hubbard, S.J., 1998. The structural aspects of limited proteolysis of native proteins. *Biochim. Biophys. Acta* 1382:191.

Itakura, K., Hirose, T., Crea, R., Riggs, A.D., Heyneker, H.L., Bolivar, F., and Boyer, H.W., 1977. Expression in *E. coli* of a chemically synthesized gene for the hormone somatostatin. *Science* 198:1056.

Jacobson, G.R., Takacs, B.J., and Rosenbusch, J.P., 1976. Properties of a major protein released from *E. coli* by osmotic shock. *Biochemistry* 15:2297.

Johnson, D.L., Middleton, S.A., McMahon, F., Barbone, F.P., Kroon, D., Tsao, E., Lee, W.H., Mulcahy, L.S., and Jollife, C.K., 1996. Refolding, purification and characterization of human erythropoietin binding protein produced in *E. coli*. *Prot. Exp. Purif.* 7:104.

Jung, G., Denefle, P., Becquart, J., and Mayaux, J.-F., 1988. High-cell density fermentation studies of recombinant Escherichia coli strains expressing human interleukin-1 beta. *Microbiol.* 139:129.

Kapust, R.B. and Waugh, D.S., 1999. *Escherichia coli* maltose-binding protein is uncommonly effective at promoting the solubility of polypeptides to which it is fused. *Protein Sci.* 8:1668.

Kendrick, B.S., Carpenter, J.F., Cleland, J.L., and Randoph, T.W., 1998. A transient expansion of the native state precedes aggregation of recombinant human interferon-gamma. *Proc. Natl. Acad. Sci. USA* 95:14142.

Khan, R.H., Appa Rao, K.B.C., Eshwari, A.N.S., Totcy, S.M., and Panda, A.K., 1998. Solubilization of recombinant ovine growth hormone with retention of native-like secondary structure and its refolding from the inclusion bodies of *E. coli*. *Biotechnol. Prog.* 14:722.

King, J., Haase-Pettingell, C., Robinson, A.S., Speed, M., and Mitraki, A., 1996. Thermolabile folding intermediates: inclusion body precursors and chaperonin substrates. *FASEB J.* 10:57.

Kita, Y., Arakawa, T., Lin, T.-Y., and Timasheff, S.N., 1994. Contribution of the surface free energy perturbation to protein-solvent interactions. *Biochemistry* 33:15178.

Krensky, A. and Nelson, P.J., 1997. Expression of chemokine RANTES and production of monoclonal antibodies. *Meth. Enzymol.* 287:162.

Koshland, D. and Botstein, D., 1980. Secretion of beta-lactamase requires the carboxy end of the protein. *Cell* 20:749.

Kudo, T., Kato, C., and Horikoshi, K., 1983. Excretion of the penicillinase of an alkalophilic Bacillus sp. through the *E. coli* outer membrane. *J. Bacteriol.* 156:949.

Kurucz, I., Titus, J.A., Jost, C.R., and Segal, D.M., 1995. Correct disulfide pairing and efficient refolding of detergent-solubilized single-chain Fv proteins from bacterial inclusion bodies. *Mol. Immun.* 32:1443.

Le, H.V. and Trotta, P.P., 1991. Purification of Secreted Recombinant Proteins from *E. coli*, in: *Purification and Analysis of Recombinant Proteins*. R. Seetharam and S.K. Sharma, eds., Marcel Dekker, New York.

Lee, J.C. and Timasheff, S.N., 1981. The stabilization of proteins by sucrose. *J. Biol. Chem.* 256:7193.

Lee, J.C. and Lee, L.L.Y., 1979. Interaction of calf brain tubulin with poly(ethylene glycols). *Biochemistry* 18:5518.

Lee, J.C. and Lee, L.L.Y., 1981. Preferential solvent interactions between proteins and polyethylene glycols. *J. Biol. Chem.* 256:625.

Libby, R.T., Bradt, G., Kronheim, S.R., March, C.J., Urdal, D.L., Chiaverotti, T.A., Tushingski, R.J., Mochizaki, D.Y., Hopp, T.P., and Cosman, D., 1987. Expression and purification of native human granulocyte-macrophage colony-stimulating factor from an Escherichia coli secretion vector. *DNA* 6:221.

Lin, T.Y. and Timasheff, S.N., 1994. Why do some organisms use a urea-methylamine mixture as osmolyte? Thermodynamic compensation of urea and trimethylamine N-oxide interactions with protein. *Biochemistry* 33:12695.

Lindwall, G., Chan, M.F., Gardner, S.R., and Kohlstaedt, L.A., 2000. A sparse matrix approach to the solubilization of overexpressed proteins. *Prot. Engineer.* 13:67.

Ma, L.-C. and Anderson, S., 1997. Correlation between disulfide reduction and conformational unfolding in bovine pancreatic trypsin inhibitor. *Biochemistry* 36:3728.

Maeda, Y., Koga, H., Yamada, H., Ueda, T., and Imoto, T., 1995. Effective renaturation of reduced lysozyme by gentle removal of urea. *Prot. Engineer.* 8:201.

Maeda, Y., Ueda, T., and Imoto, T., 1996. Effective renaturation of denatured and reduced immunoglobulin G *in vitro* without assistance of chaperone. *Prot. Engineer.* 9:95.

Martin, J., Langer, T., Boteva, R., Schramel, A., Horwich, A., and Hartl, F., 1991. Chaperonin-mediated protein folding at the surface of groEL through a "molten globule"-like intermediate. *Nature* 352:36.

Michaelis, S. and Beckwith, J., 1982. Mechanism of incorporation of cell envelope proteins in Escherichia coli. *Ann. Rev. Microbiol.* 36:435.

Mitraki, A., Betton, J.M., Desmadi, M., and Yon, J.M., 1987. Quasi-irreversibility in the unfolding-refolding transition of phosphoglycerate kinase induced by guanidine hydrochloride. *Eur. J. Biochem.* 163:29.

Mitraki, A. and King, J., 1989. Protein folding intermediates and inclusion body formation. *Bio./Technol.* 7:690.

Mitraki, A., Fane, B., Haase-Pettingell, C., Sturtevant, J., and King, J., 1991. Global suppression of protein folding defects and inclusion body formation. *Science* 253:54.

Miyake, A., Konishi, M., Martin, F.H., Hernday, N.A., Ozaki, K., Yamamoto, S., Arakawa, T., and Itoh, N., 1998. Structure and expression of a novel member, FGF-16, on the fibroblast growth factor family. *Biochem. Biophys. Res. Commun.* 243:148.

Mockman, N. and Holland, B., 1984. Secretion of a 107 K dalton polypeptide into the medium from a haemolytic *E. coli* K12 strain. *Mol. Gen. Genet.* 193:312.

Mukhopadhyay, A., 1997. Inclusion bodies and purification of proteins in biologically active forms, *in: Advances in Biochemical Engineeing/Biotechnology*, Volume 56, T. Schepe, ed.

Müller, D., Hughes, C., and Goebel, W., 1983. Relationship between plasmid and chromosomal hemolysin determinants of Escherichia coli. *J. Bacteriol.* 153:846.

Nagahari, K., Kanaya, S., Munakata, K., Aoyagi, Y., and Mizushima, S., 1985. Secretion into the culture medium of a foreign gene product from *Escherichia coli*: use of the ompF gene for secretion of human beta-endorphin. *EMBO J.* 4:3589.

Nikaido, H. and Vaaro, M., 1985. Molecular basis of bacterial outer membrane permeability. *Microbiol. Rev.* 49:1.

Nguyen, L.H. and Burgess, R.R., 1996. Overproduction and purification of sigmaS, the *Escherichia coli* stationary phase specific sigma transcription factor. *Prot. Exp. Purif.* 8:17.

Oberg, K., Chrunyk, B.A., Wetzel, R., and Fink, A.L., 1994. Nativelike secondary structure in interleukin-1 beta inclusion bodies by attenuated total reflectance FTIR. *Biochemistry* 33:2628.

Onda, M., Tatsumi, E., Takahashi, N., and Hirose, M., 1997. Refolding process of ovalbumin from urea-denatured state. Evidence for the involvement of nonproductive side chain interactions in an early intermediate. *J. Biol. Chem.* 272:3973.

Onda, M., Tatsumi, E., Takahashi, N., and Hirose, M., 1997. Refolding of urea-denatured ovalbumin that comprises non-native disulfide isomers. *J. Biochem.* 122:83.

Proudfoot, A.E., Goffin, L., Payton, M.A., Wells, T.N., and Bernard, A.R., 1996. *In vivo* and *in vitro* folding of a recombinant metalloenzyme, phosphomannose isomerase. *Biochem. J.* 318:437.

Pugsley, A.T., 1993. The complete general secretory pathway in gram-negative bacteria. *Microbiol. Rev.* 57:50.

Qu, Y., Bolen, C.L., and Bolen, D.W., 1998. Osmolyte-driven contraction of a random coil protein. *Prot. Natl. Acad. Sci.* USA 95:9268.

Ranson, N.A., White, H.E., and Saibil, H.R., 1998. Chaperonins. *Biochem. J.* 333:233.

Rao, S. and Bodley, J.W., 1996. Expression, purification, and characterization of the G domain of Saccharomyces cerevisiae elongation factor 2. *Prot. Exp. Purif.* 8:91.

Rinas, U., Risse, B., Jaenicke, R., Abel, K.J., and Zettlmeissl, G., 1990. Denaturation-renaturation of the fibrin-stabilizing factor XIII a-chain isolated from human placenta: Properties of the native and reconstituted protein. *J. Biol. Chem.* 371:49.

Ropson, I.J., Gordon, J.I., and Frieden, C., 1990. Folding of a predominantly beta-structure protein: rat intestinal fatty acid binding protein. *Biochemistry* 29:9591.

Rosenfeld, R.D., Miller, J.A., Narhi, L.O., Hawkins, N., Katta, V., Lauren, S., Weiss, M.A., and Arakawa, T., 1997. Putative folding pathway of insulin-like growth factor-I. *Arch. Biochem. Biophys.* 342:298.

Rosenfeld, R.D., Zeni, L., Welcher, A.A., Narhi, L.O., Hale, C., Marasco, J., Delaney, J., Gleason, T., Philo, J.S., Katta, V., Hui, J., Baumgartner, J., Graham, M., Stark, K.L., and Karbon, W., 1998. Biochemical, biophysical, and pharmacological characterization of bacterially expressed human agouti-related protein. *Biochemistry* 37:160–41.

Rudolph, R., Opitz, U., Hesse, F., Riebland, R., and Fischer, S., 1992. Reactivation of microbially produced human tissue-type plasminogen activator. *Biotechnol. Int.* 321–325.

Sachdev, D. and Chirgwin, J.M., 1998. Solubility of proteins isolated from inclusion bodies is enhanced by fusion to maltose-binding protein or thioredoxin. *Prot. Exp. Purif.* 12:122.

Sarmientos, P., Duchesne, M., Denefle, P., Boizian, J.B., Fromage, N., Delporte, N., Parker, F., Lelievre, Y., Mayaux, J.F., and Cartwright, T., 1989. Synthesis and purification of active human tissue plasminogen activator from escherichia-coli. *Bio/Technology* 7:495.

Sawyer, J.R., Schlom, J., and Kashmiri, S.V., 1994. The effects of induction conditions on production of a soluble anti-tumor sFv in *E. coli*. *Prot. Engineer.* 7:1401.

Schein, C.H., 1989. Production of soluble recombinant proteins in bacteria. *Bio/Technology* 7:114.

Schilstra, M.J., Slot, J.W., Van der Meide, P.H., Posthuma, G., Cremers, A.F., and Bosch, L., 1984. Immunocytochemical localization of the elongation factor Tu in *E. coli* cells. *FEBS Lett.* 165:175.

Sharma, M. et al., 1996. Expression and purification of anthrax toxin protective antigen from *E. coli*. *Prot. Exp. Purif.* 7:33.

Skerra, A. and Plückthun, A., 1988. Assembly of a functional immunoglobulin Fv fragment in *Escherichia coli*. *Science* 240:1038.

Snyder, S.W., Edalji, R.P., Lindh, F.G., Walter, K.A., Solomon, L., Pratt, S., Steffy, K., and Holzman, T.F., 1996. Initial characterization of autoprocessing and active-center mutants of CMV proteinase. *J. Prot. Chem.* 15:763.

Speed, M.A., Wang, D.I.C., and King, J., 1996. Specific aggregation of partially folded polypeptide chains: The molecular basis of inclusion body composition. *Nat. Biotechnol.* 14:1283.

Speed, M.A., Wang, D.I.C., and King, J., 1995. Multimeric intermediates in the pathway to the aggregated inclusion body state for P22 tailspike polypeptide chains. *Prot. Sci.* 4:900.

Stampher, G., Hoell-Neugeberger, B., and Rudolph, R., 1996. Improved refolding of an immobilized fusion protein. *Nature Biotechnol.* 14:329.

St. John, R.J., Carpenter, J.F., and Randolph, T.W., 1999. High pressure fosters protein refolding from aggregates at high concentration. *Proc. Natl. Acad. Sci. USA* 96:13029.

Takahara, M., Hibler, D.W., Barr, P.J., Gerlt, J.A., and Inouye, M., 1985. The ompA signal peptide directed secretion of staphylococcal nuclease A by *E. coli*. *J. Biol. Chem.* 260:2670.

Takahashi, N., Koseki, T., Doi, E., and Hirose, M., 1991. Role of an intrachain disulfide bond in the conformation and stability of ovalbumin. *J. Biochem.* 109:846.

Takahashi, N. and Hirose, M., 1992. Reversible denaturation of disulfide-reduced ovalbumin and its reoxidation generating the native cystine cross-link. *J. Biol. Chem.* 267:11565.

Takeuchi, M., Inoue, N., Strickland, T.W., Kubota, M., Wada, M., Shimizu, R., Hoshi, S., Kozutsumi, H., Takasaki, S., and Kobata, A., 1989. Relationship between sugar chain structure and biological activity of recombinant human erythropoietin produced in Chinese hamster ovary cells. *Proc. Natl. Acad. Sci. USA* 86:7819.

Talmadge, K., Stahl, S., and Gilbert, W., 1980a. Eukaryotic signal sequence transports insulin antigen in *Escherichia coli*. *Proc. Natl. Acad. Sci. USA* 77:3369.

Talmadge, K., Kaufman, J., and Gilbert, W., 1980b. Bacteria mature preproinsulin to proinsulin. *Proc. Natl. Acad. Sci. USA* 77:3988.

Tanaka, R., Tokunaga, H., Hara, S., Arakawa, T., and Tokunaga, M., 1998. Expression and purification of cytokine receptor homology domain of human granulocyte-colony stimulating factor receptor in *Escherichia coli*. *Biosci. Biotechnol. Biochem.* 62:1809.

Tatsumi, E., Takahashi, N., and Hirose, M., 1994. Denatured state of ovalbumin in high concentrations of urea as evaluated by disulfide rearrangement analysis. *J. Biol. Chem.* 269:28062.

Thomas, J.G., Ayling, A., and Baneyx, F., 1997. Molecular chaperones, folding catalysts, and the recovery of active recombinant proteins from *E. coli*. To fold or to refold. *Appl. Biochem. Biotech.* 66:197.

Tsuji, T., Nakagawa, R., Sugimoto, N., and Fukuhara K., 1987. Characterization of disulfide bonds in recombinant proteins: reduced human interleukin 2 in inclusion bodies and its oxidative refolding. *Biochemistry* 26:3129.

Völkel, D., Blankenfeld, W., and Schomburg, D., 1998. Large-scale production, purification and refolding of the full-length cellular prion protein from Syrian golden hamster in *Escherichia coli* using the glutathione S-transferase-fusion system. *Eur. J. Biochem.* 251:462.

Weissman, J.S. and Kim, P.S., 1991. Reexamination of the folding of BPTI: predominance of native intermediates. *Science* 253:1386.

Werner, M.H., Clore, G.M., Gronenborn, A.M., Kondoh, A., and Fisher, R.J., 1994. Refolding proteins by gel filtration chromatography. *FEBS Lett.* 345:125.

Wickner, W., Driessen, A.J.M., and Hartl, F.U., 1991. The enzymology of protein translocation across the Escherichia coli plasma membrane. *Annu. Rev. Biochem.* 60:101.

Wulfing, C. and Plückthun, A., 1994. Correctly folded T-cell receptor fragments in the periplasm of *E. coli*. Influence of folding catalysts. *J. Mol. Biol.* 242:655.

Yasukawa, T., Kanei-Ishii, C., Maekawa, T., Fujimoto, J., Yamamoto, T., and Ishii, S., 1995. Increase of solubility of foreign proteins in Escherichia coli by coproduction of the bacterial thioredoxin. *J. Biol. Chem.* 270:25328.

Yu, P.-L., Zhao, J., Yu, M., Reid, R., and Santi, D.V., 1996. Functional expression of the dihydrofolate reductase domain of Leishmania major dihydrofolate reductase-thymidylate synthase bifunctional protein. *Prot. Exp. Purif.* 8:23.

Zhang, J.-G., Matthews, J.M., Ward, L.W., and Simpson, R.J., 1997. Disruption of the disulfide bonds of recombinant murine interleukin-6 induces formation of a partially unfolded state. *Biochemistry* 36:2380.

Zhang, J.-G., Mortitz, R.M., Reid, G.E., Ward, L.W., and Simpson, R.J., 1992. Purification and characterization of a recombinant murine interleukin-6. Isolation of N- and C-terminally truncated forms. *Eur. J. Biochem.* 207:903.

Zhang, J.X. and Goldenberg, D.P., 1993. Amino acid replacement that eliminates kinetic traps in the folding pathway of pancreatic trypsin inhibitor. *Biochemistry* 32:14075.

3

Physical Stabilization of Proteins in Aqueous Solution

Brent S. Kendrick, Tiansheng Li, and Byeong S. Chang

INTRODUCTION

The formulation scientist's key goal is to achieve long-term stability of a drug compound. In the case of protein drugs, stabilization means not only maintaining the native chemical structure, but the native secondary and higher order structures necessary for biological activity (Cleland et al., 1993; Manning et al., 1989). Denaturation, as it is defined in this context, will be the process of forming any non-native physical or chemical state of the protein. Physical and chemical denaturations are often accompanied by covalent and non-covalent aggregates that not only can destroy the activity of the drug, but also cause adverse side effects (Carpenter and Chang, 1996; Thornton and Ballow, 1993). Without the ability to stabilize native protein structures, even the most efficacious protein therapeutics will fail to make viable drug products.

How does a formulation scientist develop a formulation that stabilizes native protein structure against physical and chemical stresses in solution, and what are the relevant stresses that cause denaturation? These are the questions that this chapter will address.

Brent S. Kendrick • Amgen, Inc., Boulder, Colorado 80301. *Tiansheng Li, and Byeong S. Chang* • Amgen, Inc. Thousand Oaks, CA 91320.

Rational Design of Stable Protein Formulations, edited by Carpenter and Manning. Kluwer Academic / Plenum Publishers, New York, 2002.

Although the chemical and physical stabilities of a protein may seem separate parameters, they are actually closely tied to one another (Brange, 1992; Khossravi et al., 2000; McCrossin et al., 1998; Rahuel-Clermont et al., 1997). Physical degradation of a protein can lead to covalent changes (oxidation, hydrolysis, disulfide scrambling). The reverse is also true; reduction of disulfide bonds, hydrolysis, and other covalent changes can cause a loss of the protein native state. For chemical degradations that are linked to physical degradation, formulations that stabilize the native state will necessarily stabilize against the chemical degradation. Understanding these relationships between physical and chemical stabilities is currently a major goal in formulation research.

OVERVIEW OF PHYSICAL STABILITY

The physical stability of a protein relates to maintaining the native secondary and higher order structures. To understand how to use formulation variables to stabilize the native structure, one must understand the mechanisms by which a protein adopts a folded conformation in solution. These mechanisms may be separated into thermodynamic and kinetic contributions to protein stability (Baker and Agard, 1994; Jaenicke, 1995).

Thermodynamic Control of Protein Stability

The traditional techniques of equilibrium unfolding have led to a vast amount of information on the thermodynamics involved in protein stability (Pace, 1975; Privalov, 1979; Schellman, 1987; Tanford, 1968). Equilibrium unfolding is typically accomplished by varying temperature or pH, or by adding denaturants such as guanidine hydrochloride or urea (Pace, 1975). To compare the stabilities of various proteins, the energetics of the unfolding reactions are typically extrapolated to conditions where the proteins exist predominantly in the native state (neutral pH, room temperature, isotonic, etc.) These techniques have given information on the enthalpic and entropic forces that stabilize the solution native state of proteins, and have generally shown that the free energy of stabilization of the native state is typically 5–20 kcal/mol (Dill, 1990; Privalov, 1979). Further information on the forces involved in protein stability has been obtained by unfolding under varying solution conditions (i.e. different pH, salt concentrations, hydrogen-bond competing molecules, etc.). The results of these experiments, coupled with experiments on model peptides, has led to an understanding of the contribution of ionic, hydrogen bonding and hydrophobic effects role in protein stability (for a review, see (Dill, 1990)).

Hydrophobic interactions are most likely the dominant forces in stabilizing the native structures of proteins (Dill, 1990; Kauzmann, 1959; Schellman, 1987).

Dill (Dill, 1990) gives a good summary of the mechanism of the hydrophobic effect, and the reader is directed there for further information. The net result of this force is that in aqueous solutions, non-polar groups interact in a strong temperature-dependent fashion. Simple thermodynamic calculations show that this condition must lead to a stability curve with a temperature of maximal stability (usually around 0 to 25°C). This necessarily results in two temperatures where free energy of unfolding is zero (i.e. a cold denaturation temperature, and a hot denaturation temperature) (Becktel and Schellman, 1987; Privalov, 1990). Many authors have observed the interesting effect of not only cold denaturation, but of cold induced dissociation and inactivation of multimeric enzymes (Privalov, 1990). Although the formulation scientist should be aware of cold induced denaturation, it should be noted that it is rare under normal solution conditions of neutral pH and moderate salt concentration.

Equilibrium unfolding studies have been very successful in broadening our understanding of the forces stabilizing the native state of proteins. However, very few proteins, especially those being developed as therapeutics, behave as well as model proteins used in many published studies. Typically, to achieve reversible unfolding, the proteins have been fairly small (< 30000 kDa.), dilute (< 1 mg/ml), and highly charged by varying pH to enhance electrostatic repulsion (i.e. decrease protein aggregation). Although a protein under infinite dilution may seek folding into the native state to reduce free energy, more realistic solution conditions may lower free energy by forcing protein-protein interactions (aggregation). This undesirable "off-pathway" reaction takes us out of the realm of reversible two-state behavior into that of the, often irreversible, effects of kinetics on protein stability.

Kinetic Control of Protein Stability

In most solution conditions—especially at the protein concentrations often employed for therapeutic doses (e.g., >2 mg/ml)—the effects of intermolecular interactions cannot be ignored. In these cases, a simple thermodynamic two-state model does not apply. Intermolecular interactions allow many possible off-pathway (i.e. other than native to unfolded) reactions to occur:

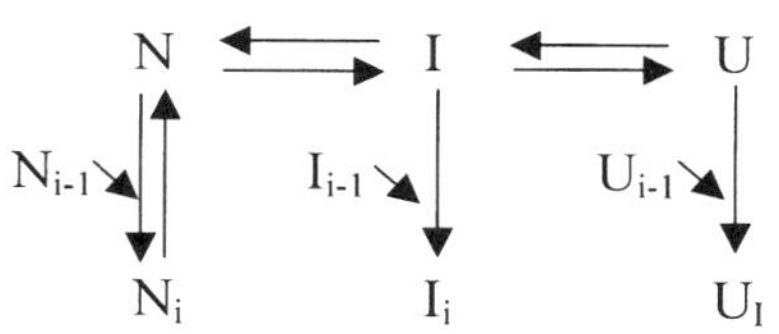

Scheme 1.

Native molecules may combine with others to form dimers or higher order native multimers. With very large N_i, crystals or amorphous precipitates can form. In most cases these transitions are reversible, because there is no or little change in protein secondary structure, and the N monomer can be formed by dilution in an appropriate solvent. The intermediate state (I) or fully unfolded (U) molecules may combine to form higher order aggregates that may either be soluble (low value of i) or insoluble (large value of i). It is now generally accepted that under non-denaturing solution conditions the pathway to I_i type aggregation is the most likely (Georgiou et al., 1994; Kendrick et al., 1998; Lumry and Eyring, 1954; Minton et al., 1982; Mulkerrin and Wetzel, 1989; Speed et al., 1996). This process is irreversible under non-denaturing conditions. In a seminal paper by Lumry and Eyring (Lumry and Eyring, 1954), irreversible aggregation was characterized by a reversible change in the native structure followed by an irreversible aggregation to the low energy intermolecular β-sheet structure. This scheme was recently modified to include a transiently expanded conformation (Kendrick et al., 1998):

$$N \rightleftharpoons N^* \longrightarrow A; \qquad A + A_{n-1} \longrightarrow A_n$$

Scheme 2.

In this scheme, N* is a transiently expanded conformation in equilibrium with N. The rate limiting step is the isomerization of N* to the aggregate competent species A, a first order reaction. The free energy diagram is given in Scheme 3. This reaction model is based on a protein that is not locked into a state N, but fluctuates between N and N*. To envision why protein conformation fluctuates about an average native structure, it is necessary to put the term "native structure" into perspective. Privalov (Privalov, 1979) stated it elegantly: "For the effective and reliable functioning of a protein, its structure must be defined over a wide range of conditions ... (However) the system must not be too rigid: It must permit a relative displacement of the structure elements in the process of protein functioning." The very nature of proteins necessitates that they exist as dynamic structures. This has been well documented with hydrogen exchange measurements (Barksdale and Rosenberg, 1983; Englander et al., 1980; Friedman, 1994; Kendrick et al., 1997a; Wang et al., 1995). Recent data suggests that under native conditions most amide groups undergo local unfolding reactions as opposed to a global cooperative unfolding fluctuation (Jeng and Englander, 1991; Todd and Freire, 1999).

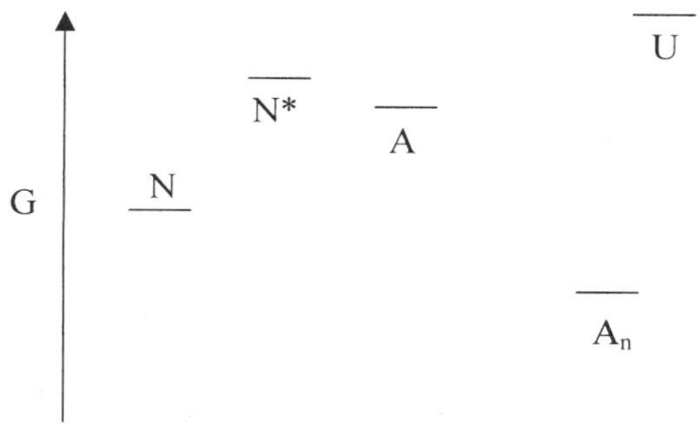

Scheme 3.

The fact that the most compact native conformation of proteins exists in reversible equilibrium with structural expanded species brings us back full circle to the application of thermodynamics to protein stability. This is particularly apparent in protein-cosolvent thermodynamics. By applying the well-documented interaction of cosolvents with proteins pioneered by Timasheff and colleagues (Timasheff, 1993), it is possible to gain a new understanding of how cosolvents (excipients) can stabilize proteins against irreversible degradation (Chang et al., 1996a; Kendrick et al., 1998; Kendrick et al., 1997a; Winzor et al., 1992; Winzor and Wills, 1995). This concept is key in the rational development of aqueous protein formulations and it will be explained briefly below. The appendix to this chapter provides a more detailed discussion on protein-cosolvent thermodynamics.

INTERACTIONS OF EXCIPIENTS WITH PROTEINS

Stabilizing proteins against physical degradation can be accomplished by choosing the appropriate excipient (also referred as cosolvent in discussions of protein-stabilizer intreractions) to target a specific stress (i.e. nonionic surfactant protection against interfacial stresses) or by choosing an excipient that acts as a general protein stabilizer. Normally excipients that act as general protein stabilizers are needed in fairly high concentrations to give a significant stabilization (> 0.25 M), and may not be suitable in cases where isotonicity must be maintained (e.g., subcutaneous doses). However, in many cases hypertonic solutions are acceptable, especially if they are diluted (e.g., into an intravenous bag), or given as a small volume intravenous injection.

Preferentially Excluded Cosolvents

Understanding the mechanism of protein stabilization by high concentrations of preferentially excluded cosolvents (excipients) is a major area of research. Timasheff and colleagues have pioneered this effort (Timasheff, 1993), using multicomponent theory (see appendix to this chapter and (Casassa and Eisenberg, 1964)), and have placed these stabilizing cosolvents into two categories. In the first, the interaction is totally independent of the chemical nature of the protein surface, i.e. it only depends on the amount of exposed surface area. For protein stabilizers, examples of these are sugars, most nonhydrophobic amino acids, and potent salting-out salts such as $(NH_4)_2SO_4$. In the second category are the cosolvents that act in a manner dependent on the chemical nature of the protein surface and the cosolvents. Examples of these types of cosolvents are polyethylene glycols, certain polyols (e.g., ethylene glycol), and alcohols. For these cosolvents, it is difficult to predict their effect on protein stability, especially at elevated temperatures (i.e., greater than room temperature). These cosolvents have substantial hydrophobic character that can foster greater binding with protein molecules, especially the denatured state, as temperature is increased. The result is that a compound (e.g., ethanol) may be a protein stabilizer at low temperature, but favor the denatured state at higher temperatures.

In terms of stabilizing interactions with proteins sucrose is the most studied cosolvent/excipient (Kendrick et al., 1998; Kendrick et al., 1997a; Lee and Timasheff, 1981; Liu and Bolen, 1995; Wang et al., 1995). It is a good model to explain the protective effect of the first category of preferentially excluded cosolvents, those for which the degree of preferential exclusion is a direct function of protein surface area. The interaction is thermodynamically unfavorable, primarily due to sucrose's repulsion from the protein backbone (Liu and Bolen, 1995), resulting in an increase in protein chemical potential (see Appendix). Thus, in the presence of sucrose, the protein state with the least surface area will be thermodynamically favored; sucrose stabilizes proteins by driving them towards a compact native state. As a result sucrose will stabilize the protein against any stress that perturbs the native conformation towards a more expanded state. In Scheme 1, in the presence of sucrose the equilibrium would be shifted towards N and away from I or U. In Schemes 2 and 3, the ΔG of the N to N* transition would increase in the presence of sucrose. As a result the level of N* would be reduced and the aggregation would be inhibited. Thus, thermodynamic stabilization of the native state can increase the kinetic stability of a protein.

It is through the preferential exclusion mechanism that sucrose and other excluded excipients stabilize proteins against thermal unfolding (Lee and Timasheff, 1981; Santoro et al., 1992), denaturant unfolding (Foord and Leatherbarrow, 1998), aggregation (Kendrick et al., 1997b), and random conformational fluctuations from the native state (Foord and Leatherbarrow, 1998;

Kendrick et al., 1997a; Wang et al., 1995). Excluded excipients may also reduce the rates chemical degradation reactions that are coupled to protein conformation, e.g. they have the ability to reduce the chemical reactivity of buried side chains, such as cysteine (Kapoor and Parfett, 1977; Kendrick et al., 1997a; Yancey and Somero, 1979).

Buffers/Salts

The choice of a buffer may not depend only on having the appropriate pK_a for the formulation. Buffers with similar pK_as can have profoundly different effects on protein stability, depending on the buffer's interaction with the protein. Salts and buffers can interact with proteins through three mechanisms: Changing the enthalpy of ionization of various side chains, a cosolvent exclusion mechanism, and a Debye screening of charge fluctuations.

Preferential Exclusion of Salts. Many buffers and/or salts at high concentrations can also stabilize proteins through the preferential exclusion mechanism. The ranking in effectiveness of stabilization follows the well-known Hofmeister series for anions (Collins and Washabaugh, 1985; Hofmeister, 1888): citrate^{3-}/citrate^{2-} > $PO_4^{3-} \cong HPO_4^{2-} \cong SO_4^{2-}$ > OAc^-, F^- > Cl^- > Br^- > I^- > ClO_4^-. Ions above Br^- in this series have an increasing ability to stabilize proteins (Jensen et al., 1995; von Hippel and Schleich, 1969; von Hippel and Wong, 1965). For example, the T_m of ribonuclease A increases by 8°C in 0.5 M potassium phosphate, relative to a solution without phosphate (von Hippel and Wong, 1965). The β-lactoglobulin dimer is similarly stabilized against urea and temperature-induced denaturation by salts in the trend predicted by the Hofmeister series (Jensen et al., 1995; Kella and Kinsella, 1988). Interestingly, cations have little effect on protein stability relative to the anions.

Although salts can have a remarkable effect on protein stability, they must be used with care, since they can also dramatically affect protein solubility. The stabilizing salts, through the preferential exclusion mechanism, can drive the minimization of protein surface area exposed to solvent to the extreme by "salting-out" the protein (Arakawa and Timasheff, 1982). The influence of various formulation conditions on protein solubility is discussed in a review by Middaugh and Volkin (Middaugh and Volkin, 1992). It should be noted that under conditions of increasing or decreasing pH from the pI of the protein, solubility of the native state increases dramatically. However, at pH extremes, the protein may unfold and irreversibly precipitate from solution through the formation of non-native intermolecular β-sheet structure.

Debye Screening. Even salts that interact weakly through the preferential exclusion mechanism, or salts in relatively low concentrations (< 0.15 M) can have a strong influence on the physical state of the protein. This is primarily due to the effects of salts on electrostatic interactions between protein molecules that are governed by charge fluctuations. Charge fluctuations in proteins are related to configurational fluctuations (Kirkwood and Shumaker, 1952a). In salt free solutions, especially those close to the isoionic point of the protein, fluctuations of charge produce a long-range attractive force with diminishing as $1/r^2$ (Kirkwood and Shumaker, 1952b). Under this solution condition, protein solubility is very small. As salt is added, the charge fluctuations contribute to a binding interaction between the protein and the salt ions (Kirkwood and Shumaker, 1952b; Scatchard, 1943). This charge screening interferes with the attractive force, and the net result is a significant increase in protein solubility (Melander and Horvath, 1977).

Ionization Enthalpy. The free energy of unfolding is by definition affected by an enthalpic transition. Buffer choice can have a strong influence on the magnitude of the enthalpy of unfolding. The mechanism for this is rooted in the cause of pH induced unfolding of a protein. The dependence of the unfolding temperature on pH is determined by the number of groups ionized upon denaturation, $\Delta_N^D\gamma$, through the relation (Privalov, 1990):

$$\Delta_N^D\gamma = -\frac{\Delta H(T_m)}{2.3T^2}\cdot\frac{dT_m}{dpH} \tag{1}$$

ΔH is the enthalpy of the transition from the native state, N, to the denatured state, D. T is temperature (in Kelvin, K) T_m is the temperature (K) of the transition midpoint from the native state to the denatured state, The enthalpy of the unfolding reaction, and thus the free energy of the transition (see (Todd et al., 1998)), is strongly dependent on the choice of buffer, since the ionization of the buffer occurs in conjunction with the ionization of protein side groups. For example, the T_m for myoglobin is 59.2° C in piperazine buffer but 73.7° C in acetate buffer at the same pH (Privalov et al., 1986). In a pharmaceutically relevant study, it was found that interferon-gamma degraded faster in succinate buffer than in acetate buffer at the same pH (Lam et al., 1997).

Specific Binding of Ligands

Protein-ligand interactions by their very nature have long been known to affect the free energy of the native state through thermodynamic linkage functions (Wyman, 1964; Wyman and Gill, 1990). Linkage functions simply state that if a protein in its native state binds a ligand with more affinity than it would in a denatured state, in the presence of the ligand the free energy of the native state is decreased more than free energy of the denatured state (see Appendix). As a result the free energy barrier between native and denatured states is increased; the native state is stabilized. More specifically, for a protein with a single binding site in the presence of a ligand L, the free energy of stabilization (ΔG_b) is given by:

$$\Delta G_b = \Delta G^0 - RT \ln \frac{(1 + K_n[L])}{(1 + K_d[L])} \tag{2}$$

where K_n is the binding constant to the native state, K_d is the binding constant to the denatured state, R is the gas constant, T is the absolute temperature, $\Delta G°$ is the free energy change of the transition in the absence of ligand, and [L] is the *free* ligand concentration (Freire, 1999; Schellman, 1987). The increase in T_m can be approximated by (Schellman, 1975):

$$T_m = \frac{T^0 \Delta H^0}{\Delta H^0 + RT^0 \ln\left(\frac{1 + K_d[L]}{1 + K_n[L]}\right)} \tag{3}$$

Where T° and $\Delta H°$ are the temperature and enthalpy of the transition in the absence of ligand. As an example, let the ligand bind only to the native state ($K_n = 10^6\,M^{-1}$, 298 K). At a free ligand concentration of 28 μM, the free energy of stabilization is 2 kcal/mol, and the corresponding increase in T_m is approximately 18°C (Figure 1).

A few applications of ligand binding will illustrate the large increases in protein stability gained by the use of ligands. One example is the use of polyanions in the formulation of acidic fibroblast growth factor (aFGF) (Tsai et al., 1993), a protein that naturally binds heparin. Although aFGF is normally unstable and aggregates extensively after only weeks of storage at 4° C, in the presence of heparin it is stable at this temperature for at least one year. Another example is the use of divalent cation metals in proteins with binding sites for such ligands. In the presence of calcium, the increase in free energy of unfolding for βγ-crystallin homolog spherulin 3a is 13.9 kcal/mol, with a corresponding T_m increase of 20°C (Kretschmar et al., 1999). Protein-receptor complexes are also stabilized through binding interactions (Li et al., 1997; Li et al., 1998).

Protein Self-Stabilization

Proteins that undergo dimerization or higher-order reversible association are stabilized by such interactions. In low molecular weight proteins this is especially apparent, with most of the stabilization against unfolding arising from interactions at the interface between subunits (Johnson and Freire, 1996; Todd

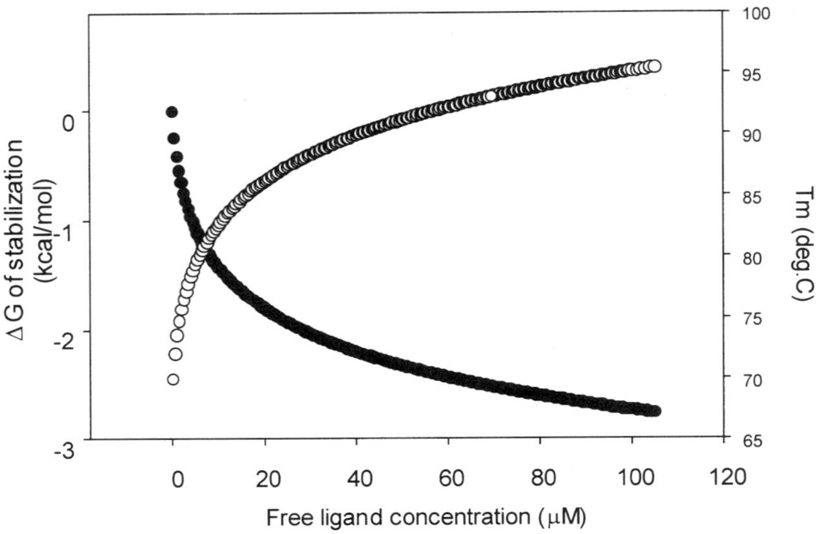

Figure 1. Hypothetical free energy of stabilization (filled circles) and corresponding increase in Tm (open circles) due to ligand binding to the native state ($K_n = 10^6 M^{-1}$).

et al., 1998). For example, in the simplest case where a protein dimer unfolds with no intermediates the stabilization against unfolding is given directly through the free energy equation:

$$\Delta G = -RT \ln K_d \qquad N_2 \xleftrightarrow{K_d} 2U \tag{4}$$

A striking example of the increase in apparent stability through protein association is seen from the protein p53tet, which has a native tetramer:unfolded monomer equilibrium in solution. As protein concentration is increased from 0.5 to 150 μM, the T_m increases from about 39 to 75° C (Johnson and Freire, 1996). Although significant gains in protein stability can be realized by increasing protein concentration, the therapeutic dose of the multimeric protein is going to have some influence on acceptable protein concentrations in the formulation.

PHYSICAL FACTORS AFFECTING PROTEIN STABILITY

Temperature

Increasing temperature will usually result in physical denaturation, chemical degradation, or both. Although thermally induced denaturation may be reversible for some proteins, most suffer irreversible denaturation and aggregation at high temperatures. The reversibility of unfolding can be strongly affected by modifying solution conditions as discussed above. Recently, a study has correlated thermal unfolding reversibility with increased long-term storage stability of a therapeutic protein against aggregation (Remmele et al., 1999). Classes of molecules that generally stabilize against thermally denaturation are ligands (Hargrove and Olson, 1996; Li et al., 1998) and preferentially excluded excipients (Rishi et al., 1998). Choice of pH, buffer, and protein concentration (for associating proteins) will also affect stability against temperature-induced denaturation.

Increasing temperature will also accelerate chemical degradation. However, at present there is no clear way to predict which type of chemical degradation will dominate at a given temperature. For example, in interleukin-1β, cysteine oxidation is the dominant chemical degradation above 38° C, but below 30° C deamidation dominates (Gu et al., 1991). This behavior brings into the question of the value of accelerated storage stability studies with respect to the investigation of chemical degradation.

Freeze-Thawing

Some proteins are quite labile to freeze-thaw stresses, resulting in extensive aggregation (Chang et al., 1996b; Jameel et al., 1997; Papandreou and Fenouillet, 1997), though the same process may not have as much negative effect on other proteins (Katakam and Banga, 1997). The molecular mechanism of freeze-thaw leading to protein aggregation remains to be fully understood. It has been postulated that freeze-thaw induced protein aggregation may be due to interfacial adsorption at ice-liquid interfaces (Kreilgaard et al., 1998; Strambini, 1996). Very limited information is known about structural changes in proteins undergoing aggregation upon freeze-thawing. In many cases, proteins can be stabilized against freeze-thawing stresses by the addition of preferentially excluded excipients or small amounts of surfactant (Chang et al., 1996b). A more detailed of stabilization during freeze-thawing is given in Chapter 7.

Agitation and Exposure to Denaturing Interfaces

During product shipping and handling, proteins in aqueous solution formulations are subjected to agitation or shaking, which is one of the most common physical stress that can cause protein aggregation (Henson et al., 1970); Maa & Hsu, 1997). Agitation induced protein denaturation and/or aggregation is presumably the consequence of protein interactions with surfaces; e.g., the air-water and vial-water interfaces (Charman et al., 1993; Henson et al., 1970; Kreilgaard et al., 1998; MacRitchie, 1986; Sluzky et al., 1991). It is generally thought that protein unfolding at air-liquid interface may irreversibly expose the interior hydrophobic core, which may in turn lead to intermolecular association of non-polar residues. It has been observed that agitation induced protein aggregation decreases at higher protein concentrations. Small amounts of surfactants are usually effective in stabilizing proteins against this type of stress (Chang et al., 1996b; Kreilgaard et al., 1998). A more detailed explanation of surface-induced protein denaturation and aggregation, and the inhibition of these processes by surfactants is given in Chapter 7.

Pressure

Although pressure-induced protein denaturation may be of least concern to protein formulation, the effect of pressure on protein stability should be considered in certain delivery devices and certain manufacturing process. In the last several years, there is an increased interest in the study of pressure denaturation of proteins. A few reported studies have shown that pressure of a few hundred MPa can induce unfolding intermediates or molten globules in protein solutions (Clery et al., 1995; Muller et al., 1981; Wong and Heremans, 1988; Yamasaki et al., 1998). Often it was found that pressure induced protein denaturation is reversible within certain range of pressure limits. It has been shown that unlike thermally denatured proteins, proteins denatured by pressure can retain their compact structures (Zhang et al., 1995). It has been proposed that pressure induced denaturation may result from the water transfer into the hydrophobic core and that the destabilization of the hydrophobic core by the penetrating water molecule leads to protein denaturation (Hummer et al., 1998; Wolfenden and Radzicka, 1994). The penetration of water molecule into protein hydrophobic core could loosen the compactness of protein interior and causes protein to swell under high pressure. Also, in general any increase in the surface area of hydrophobic residues exposed to solvent will be favored under pressure. This is because pressure favors reactions that reduce the overall system volume. Exposure of hydrophobic residues to solvent reduces volume because the density of water near these groups is greater than that of water in bulk solution. Protein structural expansion under pressure is documented by the experimental observation that lysozyme, Arc repressor, and RNase exhibit increased hydrodynamic radius upon pressure denaturation and higher hydrogen-exchange rates (Chryssomallis et al., 1981; Clery et al., 1995; Peng et al., 1994). Therefore, preferentially excluded cosolvents such as sugars and polyols will stabilize proteins against pressure-induced denaturation (Oliveira et al., 1994).

CONCLUSIONS

Long-term storage stability of proteins can be achieved through understanding of protein degradation pathways and how excipients interact with proteins. This chapter has covered many aspects of these concepts, but is by no means a comprehensive compendium. The reader is directed to other chapters in this book, resources such as this series in Pharmaceutical Biotechnology (Ahern and Manning, 1992; Pearlman and Wang, 1996; Wang and Pearlman, 1993) and the review article by Cleland et al. (Cleland et al., 1993). It is hoped that the pharmaceutical scientist can take these concepts and effectively apply them towards the rational design of stable aqueous formulations.

APPENDIX: DERIVATION OF THE WYMAN LINKAGE FUNCTION AND APPLICATION TO THE TIMASHEFF PREFERENTIAL EXCLUSION MECHANISM

Let $R \rightleftarrows P$ represent a reversible reaction (equilibrium) for any two state change of a protein going from a reactant (R) to a product (P). This includes $N \rightleftarrows D$ for unfolding, or $N \rightleftarrows P_n$ for polymerization of n monomers. The time course of a reaction towards equilibrium from any perturbed state can be illustrated by a free energy diagram (Atkins, 1994):

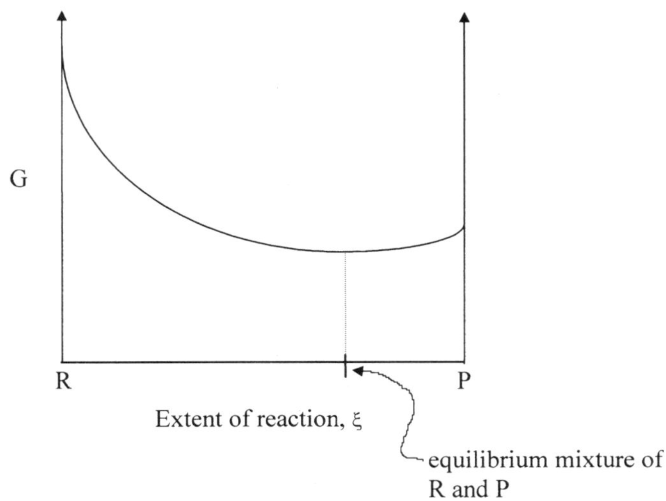

Figure a.1. Free energy of a reaction vs. extent of reaction.

For every mole P formed, one mole of R is lost. This is represented by the extent of reaction, ξ. Then, $d\xi = dn^P = -dn^R$. This relation will be useful in the interpretation of the fundamental theorem of thermodynamics (Atkins, 1994):

$$dG = Vdp - SdT + \sum_i \mu_i dn_i \tag{a1}$$

where V is volume, dp is an infinitesimal change in pressure, S is entropy, dT is an infinitesimal change in temperature, μ_I is the chemical potential (defined as $\mu_i = \partial G/\partial n_i$) of component i, and dn_i is an infinitesimal change in the number of moles of i for a given reaction. The standard notation will consist of component 1 being water, component 2 protein, and component 3 cosolvent. At constant temperature and pressure the free energy change for a reaction reduces to:

$$dG = \mu_2^P dn_2^P + \mu_2^R dn_2^R = \mu_2^P d\xi - \mu_2^R d\xi \tag{a2}$$

Taking the partial derivative at constant p and T:

$$\left(\frac{\partial G}{\partial \xi}\right)_{p,T} = \Delta G_R^P = \mu_2^P - \mu_2^R = -RT \ln K \tag{a3}$$

where K = [P]/[R].

To account for the effect of a cosolvent on the free energy of the reaction in terms of the binding parameter, $v_3 = \partial m_3/\partial m_2$ (i.e. the average statistical amount of cosolvent in the vicinity of the protein surface), take the partial derivative of the free energy equation with respect to the cosolvent (Wyman, 1964):

$$\frac{\partial(RT \ln K)}{\partial \mu_3} = \frac{\partial \mu_2^R}{\partial \mu_3} - \frac{\partial \mu_2^P}{\partial \mu_3} \tag{a4}$$

Since the definition of chemical potential depends on the activity of the solute from a given reference state, *, by:

$$\mu_3 = \mu_3^* + RT \ln(a_3) \tag{a5}$$

Then:

$$d\mu_3 = RT\, d[\ln(a_3)]$$

Substituting into equation a4 for chemical potential perturbation upon introduction of cosolvent gives the Wyman linkage function in terms of the activity of component 3:

$$\frac{\partial \ln K}{\partial \ln(a_3)} = \frac{\partial \mu_2^R}{\partial \mu_3} - \frac{\partial \mu_2^P}{\partial \mu_3} \tag{a6}$$

Dividing the top and bottom of the right hand side by ∂m_3 gives:

$$\frac{\partial \ln K}{\partial \ln(a_3)} = \left(\frac{\partial \mu_2/\partial m_3}{\partial \mu_3/\partial m_3}\right)^R - \left(\frac{\partial \mu_2/\partial m_3}{\partial \mu_3/\partial m_3}\right)^P \tag{a7}$$

Now, given a general function $z(x, y)$: $dz = \left(\frac{\partial z}{\partial x}\right)_y dx + \left(\frac{\partial z}{\partial y}\right)_x dy$ (a8a)

Take partial derivative of dz with respect to y, at constant z it will equal 0.

$$\left(\frac{\partial z}{\partial y}\right)_z = \left(\frac{\partial z}{\partial x}\right)_y \left(\frac{\partial x}{\partial y}\right)_z + \left(\frac{\partial z}{\partial y}\right)_x \left(\frac{\partial y}{\partial y}\right)_z = 0 \tag{a8b}$$

$$\Rightarrow \left(\frac{\partial z}{\partial x}\right)_y \left(\frac{\partial x}{\partial y}\right)_z = -\left(\frac{\partial z}{\partial y}\right)_x \tag{a8c}$$

Let: $\left(\frac{\partial z}{\partial y}\right)_x = \left(\frac{\partial m_3}{\partial m_2}\right)_{\mu_3}$; $\left(\frac{\partial x}{\partial y}\right)_z = \left(\frac{\partial \mu_3}{\partial m_2}\right)_{m_3} = \left(\frac{\partial \mu_2}{\partial m_3}\right)_{m_2}$; $\left(\frac{\partial z}{\partial x}\right)_y = \left(\frac{\partial m_3}{\partial \mu_3}\right)_{m_2}$

Then: $$\left(\frac{\partial \mu_2/\partial m_3}{\partial \mu_3/\partial m_3}\right) = -\left(\frac{\partial m_3}{\partial m_2}\right) \tag{a9}$$

This is the relation between binding and chemical potential perturbation.

From this relation, the Wyman linkage function becomes (Timasheff, 1993; Wyman, 1964):

$$\frac{\partial \ln K}{\partial \ln(a_3)} = \left(\frac{\partial m_3}{\partial m_2}\right)^P - \left(\frac{\partial m_3}{\partial m_2}\right)^R \tag{a10}$$

In terms of free energy:

$$\frac{\partial(\Delta G/RT)}{\partial \ln(a_3)} = \left(\frac{\partial m_3}{\partial m_2}\right)^R - \left(\frac{\partial m_3}{\partial m_2}\right)^P \tag{a11}$$

From this relationship it is apparent that a greater exclusion (negative binding) of a solute from the product (unfolded protein) than from the reactant (native protein) will result in a positive slope of free energy vs. solute concentration. Although the Wyman equation gives an indication of the direction (see graph below) that ΔG changes by the addition of a cosolvent, to obtain absolute $\Delta G^{N\text{-}D}$ values directly it is necessary to measure them by standard unfolding experiments.

It is important to note that this mechanism of salting-out and protein stabilization depends on the extent of exclusion being proportional to the exposed surface area of the protein (i.e. $\partial m_3/\partial m_2$ becomes a larger negative value as solvent exposed protein surface increases). As a technical note, it must be realized that measurement of a preferential interaction parameter for a protein in the R state under a given solution condition makes it extremely difficult to measure the preferential interaction parameter for the P state due to the direct effect of solution conditions on the equilibrium of the $R \rightleftarrows P$ transition.

Preferential interaction theory allows a direct calculation of transfer free energies of R and P from one cosolvent concentration to another. For this, knowledge of $\partial\mu_2/\partial m_3$ is necessary:

From Equation a9: $\left(\dfrac{\partial\mu_2/\partial m_3}{\partial\mu_3/\partial m_3}\right) = -\left(\dfrac{\partial m_3}{\partial m_2}\right)_{\mu_3}$

This can be rearranged to: $$(\partial_{\mu_2}/\partial m_3)_{m2} = -\frac{(\partial m_3/\partial m_2)_{\mu_3}}{(\partial\mu_3/\partial m_3)_{m_2}} \tag{a12}$$

The binding interaction, $\left(\dfrac{\partial m_3}{\partial m_2}\right)_{\mu_3}$, can be measured experimentally at equilibrium dialysis. The chemical potential perturbation of the cosolvent with itself (self interaction parameter), $\left(\dfrac{\partial\mu_3}{\partial m_3}\right)_{m_2}$, can be obtained through the equation:

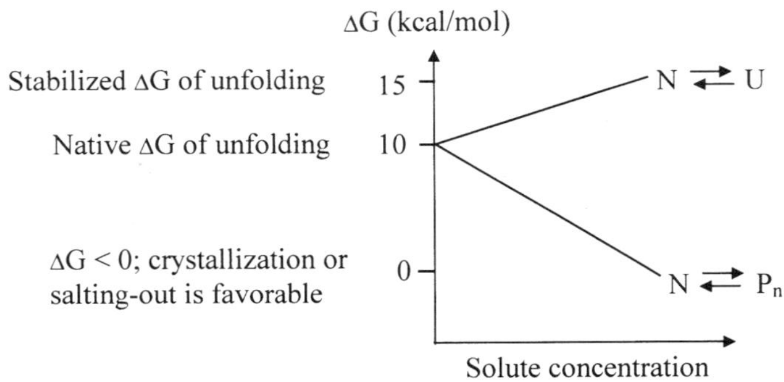

Figure a.2. Difference in ΔG ($\Delta\Delta G$) of a process upon introduction of a cosolvent.

$$\mu_3 = \mu_3^* + RT\ln(a_3) = \mu_3^* + RT\ln(m_3\gamma_3) = \mu_3^* + RT\ln(m_3) + RT\ln(\gamma_3) \qquad (a13)$$

This relationship can be transformed into:

$$\left(\frac{\partial\mu_3}{\partial m_3}\right)_{m_2} = RT/m_3 + RT\left(\frac{\partial\ln(\gamma_3)}{\partial m_3}\right)_{m_2} \qquad (a14)$$

The activity coefficient increment, $\left(\frac{\partial\ln(\gamma_3)}{\partial m_3}\right)_{m_2}$ is readily obtained from osmotic pressure data, and is usually insignificant for non-polymer cosolvents relative to RT/m_3.

The transfer free energy of R or P from water into the cosolvent system can now be obtained by (Timasheff, 1993):

$$\Delta\mu_2^R = \int_0^{m_3}\left(\frac{\partial\mu_2}{\partial m_3}\right)_{m_2}^R dm_3 \qquad (a15a)$$

$$\Delta\mu_2^P = \int_0^{m_3}\left(\frac{\partial\mu_3}{\partial m_3}\right)_{m_2}^P dm_3 \qquad (a15b)$$

Again, this parameter is difficult to obtain for U and P_n under the same solution conditions as N.

The classical "thermodynamic box" can be closed by the following relationship for N to D equilibria: $\delta\Delta G = \Delta G_{2,m_3}^{N\text{-}D} - \Delta G_{2,w}^{N\text{-}D} = \Delta\mu_{2,tr}^D - \Delta\mu_{2,tr}^N$ = free energy

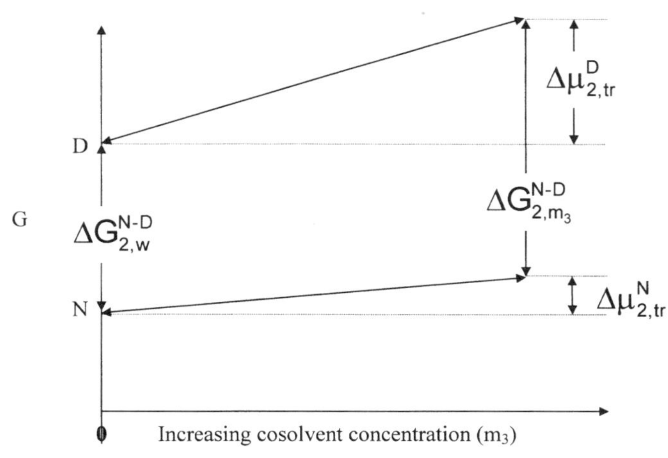

Figure a.3. Free energy of N to D transition in the absence and presence of an excluded cosolvent.

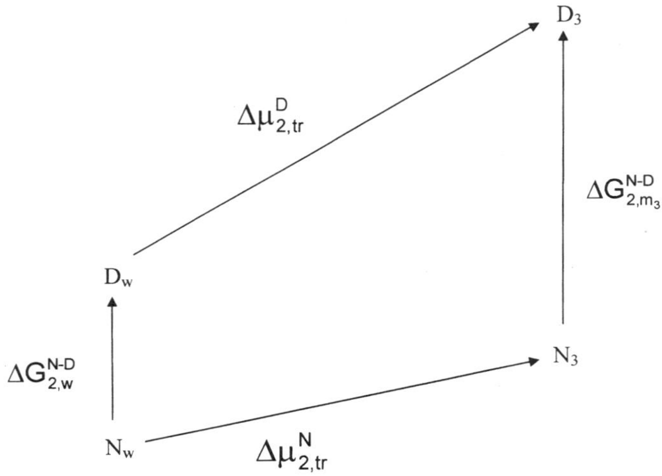

Figure a.4. Classical thermodynamic box for N to D transition in the presence and absence of an excluded cosolvent.

of stabilization due to cosolvent molality m_3 (Figures a.3 and a.4). The subscripts m_3 and w for $\Delta G^{N\text{-}D}$ refer to the change in free energy of unfolding in the presence of cosolvent (molality m_3) and the free energy of unfolding in the absence of cosolvent (e.g., in pure water or dilute buffer) respectively. These values are typically measured directly from unfolding experiments.

The second group of chemical potential perturbation terms, $\Delta\mu_{2,tr}$, refers to the free energy of transfer of D or N respectively from pure water to a cosolvent of molality m_3. This cycle has been recently completed by Timasheff's laboratory (Xie and Timasheff, 1997) for ribonuclease A in sorbitol, and can be illustrated by Figure a.4.

REFERENCES

Ahern, T.J. and Manning, M.C., 1992. Stability of protein pharmaceuticals. Plenum Press, New York.

Arakawa, T. and Timasheff, S.N., 1982. Preferential interactions of proteins with salts in concentrated solutions. *Biochemistry* 21:6545–6552.

Atkins, P.W., 1994. Physical chemistry. W. H. Freeman and Company, New York.

Baker, D. and Agard, D.A., 1994. Kinetics versus thermodynamics in protein folding. *Biochemistry* 33:7505–7509.

Barksdale, A.D. and Rosenberg, A., 1983. Aquisition and interpretation of hydrogen exchange data from peptides, polymers, and proteins. *Methods Biochem. Anal.* 28:1–113.

Becktel, W.J. and Schellman, J.A., 1987. Protein stability curves. *Biopolymers* 26:1859–1877.

Brange, J., 1992. Chemical stability of insulin. 4. Mechanisms and kinetics of chemical transformations in pharmaceutical formulation. *Acta. Pharm. Nord.* 4:209–222.

Carpenter, J.F. and Chang, B.S., 1996. Lyophilization of protein pharmaceuticals, in: *Biotechnology and Biopharmaceutical Manufacturing, Processing, and Preservation* K.E. Avis and V.L. Wu, eds., vol. 2, pp. 199–264. Interpharm. Press, Inc., Buffalo Grove, IL.

Casassa, E.F. and Eisenberg, H., 1964. Thermodynamic analysis of multicomponent systems. *Adv. Protein Chem.* 19:287–395.

Chang, B.S., Beauvais, R.M., Dong, A., and Carpenter, J.F., 1996a. Physical factors affecting the storage stability of freeze-dried interleukin-1 receptor antagonist: glass transition and protein conformation. *Arch. Biochem. Biophys.* 331:249–258.

Chang, B.S., Kendrick, B.S., and Carpenter, J.F., 1996b. Surface-induced denaturation of proteins during freezing and its inhibition by surfactants. *J. Pharm. Sci.* 85:1325–1330.

Charman, S.A., Mason, K.L., and Charman, W.N., 1993. Techniques for assessing the effects of pharmaceutical excipients on the aggregation of porcine growth hormone. *Pharm. Res.* 10:954–962.

Chryssomallis, G.S., Torgerson, P.M., Drickamer, H.G., and Weber, G., 1981. Effect of hydrostatic pressure on lysozyme and chymotrypsinogen detected by fluorescence polarization. *Biochemistry* 20:3955–3959.

Cleland, J.L., Powell, M.F., and Shire, S.J., 1993. The development of stable protein formulations: a close look at protein aggregation, deamidation, and oxidation [published erratum appears in Crit Rev Ther Drug Carrier Syst 1994;11(1):60]. *Crit. Rev. Ther. Drug. Carrier Syst.* 10:307–377.

Clery, C., Renault, F., and Masson, P., 1995. Pressure-induced molten globule state of cholinesterase. *FEBS Lett.* 370:212–214.

Collins, K.D. and Washabaugh, M.W., 1985. The Hofmeister effect and the behaviour of water at interfaces. *Quarterly Rev. Biophys.* 18:323–422.

Dill, K.A., 1990. Dominant forces in protein folding. *Biochemistry* 29:7133–7155.

Englander, S.W., Calhoun, D.B., Englander, J.J., Kallenbach, N.R., Liem, R.K., Malin, E.L., Mandal, C., and Rogero, J.R., 1980. Individual breathing reactions measured in hemoglobin by hydrogen exchange methods. *Biophys. J.* 32:577–589.

Foord, R.L. and Leatherbarrow, R.J., 1998. Effect of osmolytes on the exchange rates of backbone amide protons in proteins. *Biochemistry* 37:2969–2978.

Freire, E., 1999. The propogation of binding interactions to remote sites in proteins: Analysis of the binding of the monoclonal antibody D1.3 to lysozyme. *Proc. Natl. Acad. Sci. USA* 96:10118–10122.

Friedman, J.M., 1994. Time-resolved resonance Raman spectroscopy as probe of structure, dynamics, and reactivity in hemoglobin. *Methods Enzymol.* 232:205–231.

Georgiou, G., Valax, P., Ostermeier, M., and Horowitz, P.M., 1994. Folding and aggregation of TEM beta-lactamase: analogies with the formation of inclusion bodies in Escherichia coli. *Protein Sci.* 3:1953–1960.

Gu, L.C., Erdos, E.A., Chiang, H.S., Calderwood, T., Tsai, K., Visor, G.C., Duffy, J., Hsu, W.C., and Foster, L.C., 1991. Stability of interleukin 1 beta (IL-1 beta) in aqueous solution: analytical methods, kinetics, products, and solution formulation implications. *Pharm. Res.* 8:485–490.

Hargrove, M.S. and Olson, J.S., 1996. The stability of holomyoglobin is determined by heme affinity. *Biochemistry* 35:11310–11318.

Henson, A.F., Mitchell, J.R., and Musselwhite, P.R., 1970. The surface coagulation of proteins during shaking. *J. Colloid Interface Sci.* 32:162–165.

Hofmeister, F., 1888. On the understanding of the effect of salts. Second report. On regularities in the precipitating effect of salts and their relationship to their physiological behavior. *Naunyn-Schmiedebergs Archiv fuer Experimentelle Pathologie und Pharmakologie* 24:247–260.

Hummer, G., Garde, S., Garcia, A.E., Paulaitis, M.E., and Pratt, L.R., 1998. The pressure dependence of hydrophobic interactions is consistent with the observed pressure denaturation of proteins. *Proc. Natl. Acad. Sci. USA* 95:1552–1555.

Jaenicke, R., 1995. Folding and association versus misfolding and aggregation of proteins. *Phil. Trans. R. Soc. Lond. B.* 348:97–105.

Jameel, F., Bogner, R., Mauri, F., and Kalonia, D., 1997. Investigation of physicochemical changes to L-asparaginase during freeze-thaw cycling. *J. Pharm. Pharmacol.* 49:472–477.

Jeng, M.F. and Englander, S.W., 1991. Stable submolecular folding units in a non-compact form of cytochrome c. *J. Mol. Biol.* 221:1045–1061.

Jensen, W.A., Armstrong, J.M., DeGiorgio, J., and Hearn, M.T.W., 1995. Stability studies on maize leaf phosphoenolpyruvate carboxylase: The effect of salts. *Biochemistry* 34:472–480.

Johnson, C.R. and Freire, E., 1996. Structural stability of small oligomeric proteins. *Techniques Protein Chem.* VII:459–467.

Kapoor, M. and Parfett, C.L., 1977. Ligand-induced alterations in the reactivity of sulfhydryl groups and the structure of bovine liver glutamate dehydrogenase. *Arch. Biochem. Biophys.* 184:518–528.

Katakam, M. and Banga, A.K., 1997. Use of poloxamer polymers to stabilize recombinant human growth hormone against various processing stresses. *Pharm. Dev. Technol.* 2:143–149.

Kauzmann, W., 1959. Some factors in the interpretation of protein denaturation. *Adv. Protein Chem.* 14:1–63.

Kella, N.K.D. and Kinsella, J.E., 1988. Structural stability of β-lactoglobulin in the presence of kosmotropic salts. *Int. J. Peptide Protein Res.* 32:396–405.

Kendrick, B.S., Carpenter, J.F., Cleland, J.L., and Randolph, T.W., 1998. A transient expansion of the native state precedes aggregation of recombinant human interferon-γ. *Proc. Natl. Acad. Sci. USA* 95:14142–14146.

Kendrick, B.S., Chang, B.S., Arakawa, T., Peterson, B., Randolph, T.W., Manning, M.C., and Carpenter, J.F., 1997a. Preferential exclusion of sucrose from recombinant interleukin-1 receptor antagonist: Role in restricted conformational mobility and compaction of native state. *Proc. Natl. Acad. Sci. USA* 94:11917–11922.

Kendrick, B.S., Cleland, J.L., Lam, X., Nguyen, T., Randolph, T.W., Manning, M.C., and Carpenter, J.F., 1997b. Aggregation of recombinant human interferon gamma: Kinetics and structural transitions. *submitted to J. Pharm. Sci.*

Khossravi, M., Shire, S.J., and Borchardt, R.T., 2000. Evidence for the involvement of histidine A(12) in the aggregation and precipitation of human relaxin induced by metal-catalyzed oxidation. *Biochemistry* 39:5876–5885.

Kirkwood, J.G. and Shumaker, J.B., 1952a. *Proc. Natl. Acad. Sci. USA* 38:855–862.

Kirkwood, J.G. and Shumaker, J.B., 1952b. Forces between protein molecules in solution arising from fluctuations in proton charge and configuration. *Proc. Natl. Acad. Sci. USA* 38:863–871.

Kreilgaard, L., Jones, L.S., Randolph, T.W., Frokjaer, S., Flink, J.M., Manning, M.C., and Carpenter, J.F., 1998. Effect of Tween 20 on freeze-thawing- and agitation-induced aggregation of recombinant human factor XIII. *J. Pharm. Sci.* 87:1597–1603.

Kretschmar, M., Mayr, E.M., and Jaenicke, R., 1999. Kinetic and thermodynamic stabilization of the betagamma-crystallin homolog spherulin 3a from Physarum polycephalum by calcium binding. *J. Mol. Biol.* 289:701–705.

Lam, X.M., Patapoff, T.W., and Nguyen, T.H., 1997. The effect of benzyl alcohol on recombinant human interferon-gamma. *Pharm. Res.* 14:725–729.

Lee, J.C. and Timasheff, S.N., 1981. The stabilization of proteins by sucrose. *J. Biol. Chem.* 256:7193–7201.

Li, T., Horan, T., Osslund, T., Stearns, G., and Arakawa, T., 1997. Conformational changes in G-CSF/Receptor complex as investigated by isotope-edited FTIR spectroscopy. *Biochemistry* 36:8849–8857.

Li, T., Narhi, L.O., Wen, J., Philo, J.S., Sitney, K., Inoue, J., Yamamoto, T., and Arakawa, T., 1998. Interactions between NFkappaB and its inhibitor ikappaB: biophysical characterization of a NFkappaB/ikappaB-alpha complex. *J. Protein Chem.* 17:757–763.

Liu, Y. and Bolen, D.W., 1995. The peptide backbone plays a dominant role in protein stabilization by naturally occurring osmolytes. *Biochemistry* 34:12884–12891.

Lumry, R. and Eyring, H., 1954. Conformation changes of proteins. *J. Phys. Chem.* 58:110–120.

MacRitchie, F., 1986. Spread monolayers of proteins. *Adv. Colloid Interface Sci.* 25:341–385.

Manning, M.C., Patel, K., and Borchardt, R.T., 1989. Stability of protein pharmaceuticals. *Pharm. Res.* 6:903–918.

McCrossin, L.E., Charman, W.N., and Charman, S.A., 1998. Degradation of recombinant porcine growth hormone in the presence of guanidine hydrochloride. *Int. J. Pharm.* 173:157–170.

Melander, W. and Horvath, C., 1977. Salt effects on hydrophobic interactions in precipitation and chromatography of proteins: An interpretation of the lyotropic series. *Arch. Biochem. Biophys.* 183:200–215.

Middaugh, C.R. and Volkin, D.B., 1992. Protein solubility, in: *Stability of protein pharmaceuticals* T.J. Ahern and M.C. Manning, eds., vol. 2, pp. 109–134. Plenum Press, New York.

Minton, K.W., Karmin, P., Hahn, G.M., and Minton, A.P., 1982. Nonspecific stabilization of stress-susceptible proteins by stress-resistant proteins: a model for the biological role of heat shock proteins. *Proc. Natl. Acad. Sci. USA* 79:7107–7111.

Mulkerrin, M.G. and Wetzel, R., 1989. pH dependence of the reversible and irreversible thermal denaturation of gamma interferons. *Biochemistry* 28:6556–6561.

Muller, K., Ludemann, H.D., and Jaenicke, R., 1981. Pressure-induced structural changes of pig heart lactic dehydrogenase. *Biophys. Chem.* 14:101–110.

Oliveira, A.C., Gaspar, L.P., Da Poian, A.T., and Silva, J.L., 1994. Arc repressor will not denature under pressure in the absence of water. *J. Mol. Biol.* 240:184–187.

Pace, C.N., 1975. The stability of globular proteins. *CRC Critical Reviews in Biochemistry* 3:1–41.

Papandreou, M.J. and Fenouillet, E., 1997. Effect of various glycosidase treatments on the resistance of the HIV-1 envelope to degradation. *FEBS Lett.* 406:191–195.

Pearlman, R. and Wang, Y.J., 1996. Formulation, characterization, and stability of protein drugs. Plenum Press, New York.

Peng, X., Jonas, J., and Silva, J.L., 1994. High-pressure NMR study of the dissociation of Arc repressor. *Biochemistry* 33:8323–8329.

Privalov, P.L., 1979. Stability of proteins. *Adv. Protein Chem.* 33:167–241.

Privalov, P.L., 1990. Cold denaturation of proteins. *Crit. Rev. Biochem. Mol. Biol.* 25:281–306.

Privalov, P.L., Griko, Y.V., and Venyaminov, S.Y., 1986. Cold denaturation of myoglobin. *J. Mol. Biol.* 190:487–498.

Rahuel-Clermont, S., French, C.A., Kaarsholm, N.C., and Dunn, M.F., 1997. Mechanisms of stabilization of the insulin hexamer through allosteric ligand interactions. *Biochemistry* 36:5837–5845.

Remmele, R.L.J., Bhat, S.D., Phan, D.H., and Gombotz, W.R., 1999. Minimization of recombinant human Flt3 ligand aggregation at the Tm plateau: a matter of thermal reversibility. *Biochemistry* 38:5241–5247.

Rishi, V., Anjum, F., Ahmad, F., and Pfeil, W., 1998. Role of non-compatible osmolytes in the stabilization of proteins during heat stress. *Biochem. J.* 329:137–143.

Santoro, M.M., Liu, Y., Khan, S.M., Hou, L.X., and Bolen, D.W., 1992. Increased thermal stability of proteins in the presence of naturally occurring osmolytes. *Biochemistry* 31:5278–5283.

Scatchard, G., 1943. Chapter 3, in: *Proteins, amino acids and peptides* C.A. Edsall, eds. Reinhold, New York.

Schellman, J.A., 1975. Macromolecular binding. *Biopolymers* 14:999–1018.

Schellman, J.A., 1987. The thermodynamic stability of proteins. *Ann. Rev. Biophys. Biophys. Chem.* 16:115–137.

Sluzky, V., Tamada, J.A., Klibanov, A.M., and Langer, R., 1991. Kinetics of insulin aggregation in aqueous solutions upon agitation in the presence of hydrophobic surfaces. *Proc. Natl. Acad. Sci. USA* 88:9377–9381.

Speed, M.A., Wang, D.I.C., and King, J., 1996. Specific aggregation of partially folded polypeptide chains: The molecular basis of inclusion body composition. *Nature Biotechnology* 14:1283–1287.

Strambini, G.B.G.-E., 1996. Proteins in frozen solutions: Evidence of ice-induced partial unfolding. *Biophys. J.* 70:971–976.

Tanford, C., 1968. Protein denaturation: Part A. Characterization of the denatured state: Part B. The transition from the native to denatured state. *Adv. Protein Chem.* 23:121–275.

Thornton, C.A. and Ballow, M., 1993. Safety of intravenous immunoglobulin. *Arch. Neurol.* 50:135–136.

Timasheff, S.N., 1993. The control of protein stability and association by weak interactions with water: How do solvents affect these processes? *Annu. Rev. Biophys. Biomol. Struct.* 22:67–97.

Todd, M.J. and Freire, E., 1999. The effect of inhibitor binding on the structural stability and cooperativity of the HIV-1 protease. *Proteins* 36:147–156.

Todd, M.J., Semo, N., and Freire, E., 1998. The structural stability of the HIV-1 protease. *J. Mol. Biol.* 283:475–488.

Tsai, P.K., Volkin, D.B., Dabora, J.M., Thompson, K.C., Bruner, M.W., Gress, J.O., Matuszewska, B., Keogan, M., Bondi, J.V., and Middaugh, C.R., 1993. Formulation design of acidic fibroblast growth factor. *Pharm. Res.* 10:649–659.

von Hippel, P.H. and Schleich, T., 1969. The effects of neutral salts on the structure and conformational stability of macromolecules in solution, in: *Structure and stability of biological macromolecules* S.N. Timasheff and G.D. Fasman, eds., pp. 417–574. Marcel Dekker, New York.

von Hippel, P.H. and Wong, K.Y., 1965. On the conformational stability of globular proteins. *J. Biol. Chem.* 240:3909–3923.

Wang, A., Robertson, A.D., and Bolen, D.W., 1995. Effects of a naturally occurring compatible osmolyte on the internal dynamics of ribonuclease A. *Biochemistry* 34:15096–15104.

Wang, Y.J. and Pearlman, R., 1993. Stability and characterization of protein and peptide drugs: Case histories. Plenum Press, New York.

Winzor, C.L., Winzor, D.J., Paleg, L.G., Jones, G.P., and Naidu, B.P., 1992. Rationalization of the effects of compatible solutes on protein stability in terms of thermodynamic nonideality. *Arch. Biochem. Biophys.* 296:102–107.

Winzor, D.J. and Wills, P.R., 1995. Thermodynamic nonideality of enzyme solutions supplemented with inert solutes: yeast hexokinase revisited. *Biophys. Chem.* 57:103–110.

Wolfenden, R. and Radzicka, A., 1994. On the probability of finding a water molecule in a nonpolar cavity. *Science* 265:936–937.

Wong, P.T. and Heremans, K., 1988. Pressure effects on protein secondary structure and hydrogen deuterium exchange in chymotrypsinogen: a Fourier transform infrared spectroscopic study. *Biochim. Biophys. Acta.* 956:1–9.

Wyman, J., 1964. Linked functions and reciprocal effects in hemoglobin: A second look. *Adv. Protein Chem.* 19:223–286.

Wyman, J. and Gill, S.J., 1990. Binding and linkage: Functional chemistry of biological macromolecules. University Science Books, Mill Valley, CA.

Xie, G. and Timasheff, S.N., 1997. Mechanism of the stabilization of ribonuclease A by sorbitol: preferential hydration is greater for the denatured then for the native protein. *Protein Sci.* 6:211–221.

Yamasaki, K., Taniguchi, Y., Takeda, N., Nakano, K., Yamasaki, T., Kanaya, S., and Oobatake, M., 1998. Pressure-denatured state of Escherichia coli ribonuclease HI as monitored by Fourier transform infrared and NMR spectroscopy. *Biochemistry* 37:18001–18009.

Yancey, P.H. and Somero, G.N., 1979. Counteraction of urea destabilization of protein structure by methylamine osmoregulatory compounds of elasmobranch fishes. *Biochem. J.* 183:317–323.

Zhang, J., Peng, X., Jonas, A., and Jonas, J., 1995. NMR study of the cold, heat, and pressure unfolding of ribonuclease A. *Biochemistry* 34:8631–8641.

4

Effects of Conformation on the Chemical Stability of Pharmaceutically Relevant Polypeptides

Jeffrey D. Meyer[1], *Bert Ho*[2], *and Mark C. Manning*[1]

INTRODUCTION

Although most of the attention on the stability of protein and peptide pharmaceuticals has focused on denaturation and aggregation, chemical degradation cannot be ignored. A large number of possible chemical reactions have been identified as leading to decomposition of polypeptides (Manning et al., 1989). However, a select few are of primary concern to pharmaceutical scientists. In particular, hydrolysis and oxidation reactions appear to be the most common source of chemical instability. Hydrolysis of peptides refers to the nucleophilic addition of water to either an amide linkage in the peptide backbone (in a process known as proteolysis) or to an amide side chain of either asparagine or glutamine (known as deamidation). Meanwhile, oxidation can occur at cysteine residues (to form disulfide linkages), at methionine (to form sulfoxides), or at heterocyclic aromatic side chains of tryptophan and histidine (to form N-oxides).

Jeffrey D. Meyer and Mark C. Manning • Center for Pharmaceutical Biotechnology, Department of Pharmaceutical Sciences, School of Pharmacy, University of Colorado Health Sciences Center, Denver, CO 80262. *Bert Ho* • Tularik Corporation, S. San Francisco, CA 94080.

Rational Design of Stable Protein Formulations, edited by Carpenter and Manning. Kluwer Academic / Plenum Publishers, New York, 2002.

The mechanistic details of these reactions and the effects of extrinsic factors, such as pH, temperature, ionic strength and buffer composition have been widely reviewed (Manning et al., 1989; Nguyen, 1994; Cleland et al., 1993; Li et al., 1995; Vogt, 1995; Xie and Schowen, 1999; Aswad, 1995; Geiger and Clarke, 1987). Therefore, the focus of this chapter will be on the effects of conformation and structure on the rate of these chemical reactions. It is important to note that, in most cases, the rate of oxidation or hydrolysis of formulated protein products is relatively slow (only a small percentage per month), although this is still significant on the timescale of the typical shelf life (12–24 months). If a change in protein structure or structural stability can cause a slower rate of chemical degradation, new approaches towards stabilization of protein pharmaceuticals can be envisioned.

Previously, it had been thought that the only approach to slowing chemical degradation was to alter pH, lower the temperature, change the buffer, or, in the case of oxidation, add a scavenger. Certainly, these factors are critical in controlling chemical decomposition. However, an additional method for stabilizing proteins towards chemical degradation exists. The same strategies that are currently employed for stabilizing the native state of a protein molecule, thereby diminishing denaturation and aggregation, may also provide a benefit in retarding chemical instability. We hypothesize that both hydrolysis and oxidation reactions in proteins are, to some degree, sensitive to the overall conformation of the molecule. By modulating the structure and stability of the native protein molecule through interactions with excipients, we may be able to further slow chemical inactivation processes.

Except for some work on the deamidation of peptides and oxidation of a methionine at the active site of subtilisin (DePaz et al., 2000), this hypothesis is largely untested. Therefore, this chapter should be viewed as a proposed stabilization strategy rather than a completed demonstration of first principles. Building on what is known about conformational control of deamidation, we propose general methods of controlling other reactions, such as the oxidation of partially exposed methionine residues in proteins and peptides.

RELATIONSHIP BETWEEN STRUCTURE AND DEAMIDATION RATES

Deamidation, or hydrolysis of the side chains of Asn and Gln residues, has been extensively studied (Manning et al., 1989; Cleland et al., 1993; Xie and Schowen, 1999; Aswad, 1995; Geiger and Clarke, 1987). Except for a few isolated cases, most deamidation reactions have been found to occur at Asn rather than Gln. Presumably, this is due to the formation of a cyclic imide intermediate, the dominant mechanism for deamidation at pH values above 6 (see Figure 1). In the case of Asn, a five-membered ring is formed, whereas for Gln, a six-membered ring is necessary. The relative stability of a heterocyclic five-membered ring vs. a six-membered ring is sufficient to favor deamidation of Asn in most cases. A number of references to deamidation in pharmaceutically relevant compounds are provided to demonstrate the immense activity that has occurred in this field. For example, deamidation has been reported for epidermal growth factor (Araki et al., 1989; DiAugustine et al., 1987; Senderoff et al., 1994; George-Nascimento et al., 1990; Son and Kwon, 1995), growth hormone (Becker et al., 1988; Frenz et al., 1989; Johnson et al., 1989; Lewis et al., 1970; Nielsen et al., 1989; Pikal et al., 1991; Cleland et al., 1993, 1997; McCrossin et al., 1998), tissue plasminogen activator (Pearlman and Nguyen, 1992; Chloupek et al., 1989), insulin (Fisher and Porter, 1981; Nielsen et al., 1989; Brange et al., 1992; Berson and Yalow, 1966; Oliva et al., 1996; Strickley et al., 1996; Darrington and Anderson, 1994), monoclonal antibodies (Liu, 1992; Kroon et al., 1992; Tsai et al., 1993), interleukin-1 receptor antagonist (Chang et al., 1996), hirudin (Gietz et al., 1998), and human stem cell factor (Hsu et al., 1998).

Primary Structure Effects

Considering the mechanism shown in Figure 1, one would expect that there would be an effect of primary structure on deamidation rates, simply based upon steric hindrance. If the rate-limiting step at neutral or basic pH is nucleophilic attack to form the cyclic imide (also called succinimide) intermediate, then increasing the bulk of the adjacent amino acid side chain should slow the reaction. Conversely, if the succeeding side chain can hydrogen bond to the amide side chain of Asn (as in the case of Ser and Thr), it would make the amide more electrophilic and the reaction might proceed faster than expected.

The influence of primary structure was examined in studies by Geiger and Clarke (1987) on a hexapeptide sequence based on residues from the peptide hormone, ACTH. These studies indicated that deamidation under physiological conditions proceeds through the succinimide intermediate and the presence of a bulky residue (Leu or Pro) on the N-terminal side of the Asn residue can reduce the relative rate of degradation by a factor of 33–50 fold. In addition, Geiger and Clarke (1987) proposed that the conformational flexibility of the short peptides allows the peptide to assume the dihedral angles necessary to form the cyclic intermediate. This conformational flexibility may not be as readily achieved within the more highly ordered structure of proteins and larger peptides.

In similar studies examining the effect of primary sequence on asparagine deamidation, Patel and Borchardt (1990) varied the amino acid residue on either

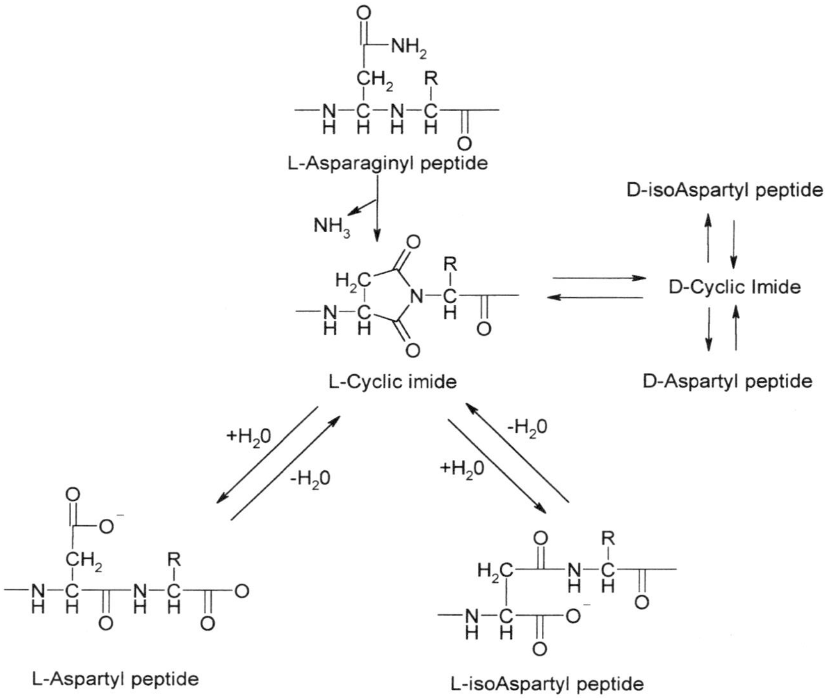

Figure 1. General mechanistic scheme for deamidation of asparagine (Asn) residues in peptide and proteins.

side of Asn in the same hexapeptide sequence, Val-Tyr-X-Asn-Y-Ala. They found that in acidic media, the Asn residue deamidates to give only Asp and there is no effect from the neighboring amino acid. This should not be surprising, as the acid-catalyzed reaction does not proceed through the cyclic imide intermediate, but rather by direct hydrolysis of the Asn side chain. At neutral or basic pH, both Asp and isoAsp products were found, following the mechanism outlined in Figure 1. Deamidation proceeded most rapidly with a Gly residue on the C-terminus side of Asn; the rate with Asn-Gly hexapeptide was 120–180 times faster than that of the more sterically-hindered Asn-Val hexapeptide. Modifications on the N-terminal side of Asn did not have any effect on deamidation at neutral or basic pH.

Tyler-Cross and Schirch (1991) examined the influence of the neighboring residue on deamidation in the pentapeptide sequences, Val-X-Asn-Ser-Val or Val-Ser-Asn-X-Val. Their results are similar to those of Patel and Borchardt (1990). They found minimal influence from the residue on the N-terminal side at any pH, and at neutral or basic pH they found the sequences with bulky residues on the

C-terminal side had the slowest rate of deamidation. Consistent with a deamidation mechanism via a succinimide intermediate at neutral or alkaline pHs, the ratio of isoAsp to Asp product was found to be consistently about 3 : 1.

Brennan and Clarke (1995) investigated the influence of adjacent basic amino acids on Asn deamidation via the succinimide intermediate. In studies with hexapeptides, they found minimal effect on succinimide formation from His or Cys residues on the C-terminal side of the Asn residue and from His on the N-terminal side of Asn. However, they did find enhancement of the competing peptide bond cleavage with His or Cys on the C-terminus side. Whether these findings apply to similar peptide sequences in mutant hemoglobins that are susceptible to deamidation is not clear as higher order structural elements are present in folded proteins.

In a theoretical study, Wright (1990) examined the influence of the nearest neighboring residue on Asn deamidation in known protein structures. The data indicate that potential "catalytic" residues neighboring Asn are selected against in a large number of proteins. Catalytic residues are those that could hydrogen bond to the neighboring Asn side chain, and render it more susceptible to intramolecular cyclization. In proteins where deamidation has been characterized, the participation of catalytic residues was made possible by the secondary or tertiary structure of the protein; the higher-order flexible conformation brings catalytic amino acids in proximity to the labile Asn residue.

These studies on short model peptides clearly show the influence of primary structure on the deamidation of Asn residues. The influence of neighboring steric factors is clear and was expected. Less understood is the effect of conformation on deamidation in peptides and proteins. One would expect higher order structure to affect the proximity of functional groups relative to the labile Asn residue, thus impacting the rate of formation of the cyclic intermediate. Higher order structure could also inhibit or accelerate the rate of deamidation by influencing the local environment of the Asn residue and the stability of transition state intermediates in the deamidation pathway, as described in the sections below.

Secondary Structure Effects

The influence of secondary structural environments of peptides and proteins on deamidation of asparagine residues has recently been reviewed by Xie and Schowen (1999). In general, the conformation adopted by the polypeptide backbone, as well as the side chains, affects the ability for intramolecular cyclization. Since the formation of the cyclic imide is the rate-controlling step, different secondary structures should, and do, exhibit different rates of deamidation, independent of primary structure effects. In an early theoretical study, Clarke (1989) examined the sequence and conformation of Asp and Asn residues in proteins with known three-dimensional structures. The optimal set of dihedral angles for cyclic imide formation as well as those least favoring cyclization are described. The findings suggest that primary sequence may not be the major determinant of deamidation in any of the stable proteins studied, rather the Asn and Asp residues in stable proteins are constrained by their conformation from attack by neighboring residues. Differences in deamidation rates result when changes in protein conformation allow local flexibility in the polypeptide chain, thus allowing formation of the imide intermediate necessary for deamidation.

The degradation of Asp and Asn residues in human growth hormone-releasing factor (GRF) was studied by Bongers et al. (1992). In GRF, Asn-8 deamidates to iso-Asp and Asp in a ratio of about 4 : 1 at neutral and alkaline pH, and exclusively to Asp at pH ~ 2. No evidence of deamidation at Asn-35 was found, possibly due to steric constraints imposed by an adjacent bulky Gln-36 residue. Deamidation of Asn-8 is general base-catalyzed and proceeds via the cyclic imide intermediate. Since deamidation of Asn-8 was found to obey non-linear Arrhenius kinetics, this suggests that a distinct temperature-dependent change in conformation has occurred, possibly conversion between an α-helical conformation and a more unordered structure.

Stevenson et al. (1993a,b) studied the effect of secondary structure on deamidation at Asn-8 with GRF analogs modified at position 15 with Pro (helix breaker) or Ala (helix enhancer). They showed the more α-helical Ala-15 analog had a decreased rate of deamidation in aqueous methanol solutions whereas the Pro-15 analog with less helical propensity had an increased rate of deamidation and less bioactivity. The increased stability of the Ala-15 analog would appear to conflict with results with a similar Ala-15 GRF analog reported by Bongers et al. (1992). Whether the presence of methanol in the later study affects the results is unclear.

Lura and Schirch (1988) compared the deamidation of the tetrapeptides Val-Asn-Gly-Ala and N-acetyl-Val-Asn-Gly-Ala. At pH < 7.5, the N-acetyl peptide undergoes deamidation of Asn to give the isoAsp and Asp products in a 3 : 1 ratio via the succinimide pathway. Surprisingly, the non-acetylated peptide gave rise to a 7-member cyclic peptide formed from the displacement of the Asn side-chain amide by the free amino terminus of the Val residue. At basic pH, both tetrapeptides deamidated by the same mechanism to give the isoAsp and Asp products. The authors proposed that the difference in deamidation rates and mechanism can be attributed to variation in the solution conformation of the peptide segment, but no structural information was provided.

The influence of other secondary structures has also been studied in proteins. Sharma et al. (1993) examined the deamidation of histidine-containing protein, HPr, in *E. coli*. Their findings indicated that HPr undergoes sequential deamidation of Asn-38 and then Asn-12 to yield, respectively, HPr-1 and HPr-2. Both Asn residues are followed by a Gly, and both deamidation reactions proceed via the cyclic imide at neutral and alkaline pH. The more labile Asn-38 is in a type III β-turn structure and may be solvent-exposed. The less susceptible Asn-12 is in a type I β-turn. This residue is stable to deamidation under normal conditions but may be susceptible once the tertiary structure is altered by the formation of the deamidated HPr-1 species.

Tertiary Structure Effects

Kossiakoff (1988) examined the factors influencing the deamidation of trypsin using a database derived from the neutron crystallographic structure of trypsin. He concluded that contrary to other findings, the steric or electronic nature of the residues flanking Asn is not the primary factor in the rate of deamidation. Modeling suggests the influence of tertiary structure in achieving or constraining the required geometry about the labile Asn residue for deamidation.

Chen et al. (1999) looked at the effect of removal of exogenous calcium ion on recombinant human DNase I (rhDNase) on deamidation and aggregation. Their findings indicate the removal of Ca^{2+} by EGTA treatment had little effect on secondary or tertiary structure of the protein. However, the removal of Ca^{2+} did result in increased rates of deamidation and soluble aggregation. It is believed that Ca^{2+} bound to the native protein protects cystine bridges against reduction; the removal of calcium ions would allow the cleavage of disulfide bridge Cys173–Cys209, thus allowing increased conformational freedom which would lead to increased susceptibility to deamidation via the succinimide intermediate. Here we have an example of metal binding apparently modulating the conformational flexibility of the protein. In the absence of metal binding, there is increased flexibility, allowing intramolecular nucleophilic attack to proceed, which leads to increased levels of chemical degradation.

Other evidence for the effect of conformation on deamidation is provided in studies on intact proteins. Artigues et al. (1990) examined deamidation of Asn in rabbit liver cytosolic serine hydroxymethyltransferase (SHMT). The rates of deamidation of Asn-5 in the intact protein was compared with that of the same residue in the cleaved 14-mer from the N-terminus of SHMT. The rate of deamidation in the intact enzyme was found to be 5-fold slower than that in the peptide fragment. This difference in rate can be attributed to conformational differences in which the Asn residue is stabilized by higher-order structure in the intact protein, whereas it is not protected in the peptide fragment. Violand et al. (1990, 1992) examined the deamidation of both Asn-99 and Asn-129 in bovine somatotropin (BST). Their results led them to conclude that the three-dimensional conformation of BST places the Asn residues in a flexible portion of the molecule, allowing the more facile formation of the succinimide intermediate.

Brange et al. (1992) examined the degradation of several formulations of insulin under various conditions. At neutral pH, Asn-B3 was found to deamidate to a mixture of Asp and iso-Asp products. The lowest rates of deamidation were seen in formulations containing phenol. This effect was attributed to the stabilizing effect of phenol on the α-helical conformation of the protein, thus reducing the possibility of cyclic imide formation. Differences in the rate of deamidation of insulin in amorphous versus crystalline suspensions were also found; the amorphous suspensions had a greater rate of deamidation than the crystalline product. Again, the authors propose the more tightly packed crystalline suspension constrains the tertiary structure, thus reducing the flexibility required to form the required cyclic imide intermediate.

In a study comparing the rates of deamidation of Asn-67 in native and unfolded RNase A, Wearne and Creighton (1989) found the native conformation inhibited deamidation of Asn-67 by 30-fold. The native conformation does not achieve the structure necessary for the formation of the cyclic imide intermediate whereas the more flexible unfolded molecule is not restricted in this region. Di Donato et al. (1986) examined the deamidation of Asn-67 in RNase A to Asp and isoAsp under mild conditions. They observed the deamidation of Asn-67 in bovine serum RNase (bsRNase) proceeded at a much faster rate yet the homology between RNase A and bsRNase is greater than 80% and the sequence about Asn-67 is identical except for residue 64. They hypothesized a role for bsRNase dimeric structure in determining the deamidation rate of Asn-67.

Summary of Structure Effects on Deamidation

Primary structure clearly affects the rate of deamidation in peptides and proteins. Extensive literature documents the influence of secondary, tertiary and even quarternary structure on the deamidation of pharmaceutically relevant products and native proteins. Many of these reports also examine the influence of the altered residue on the structure and function of the "degraded" product. Control of conformation is highly desirable in protein formulations to minimize physical degradation. These examples of structural influence on a particular form of chemical degradation, deamidation, suggest that conformational control may also aid in minimizing certain aspects of chemical degradation.

ROLE OF STRUCTURE IN PROTEIN OXIDATION

Along with deamidation, oxidation is the most predominant form of chemical degradation of pharmaceutical proteins and peptides (Li et al., 1995a; Cleland et al., 1993). There have been several excellent reviews which detail general protein oxidation (Vogt, 1995), oxidation of pharmaceutical dosage forms of proteins (Manning et al., 1989; Nguyen, 1994; Kosen, 1992), and strategies for minimizing oxidation in pharmaceuticals (Li et al., 1995a). From these reviews, a detailed understanding of the mechanisms of oxidation and some strategies for reducing oxidative rates can be gleaned. Primary strategies for minimizing oxidation include reducing the levels of reactive species (free radicals), altering the solution pH, changing the buffer species, reducing levels of molecular O_2, or the addition of a scavenger or oxidative "sponge".

As with deamidation, it may be possible to extend the use of conformational stabilization beyond physical stability issues to the realm of chemical stability, and specifically to oxidation. At this time, this hypothesis remains largely untested, and the following discussion will examine how to apply what is currently known about protein oxidation to develop a new strategy for reducing oxidative damage in polypeptides using conformational control.

Types of Oxidation Processes

In order to reduce oxidation by the control of protein conformation, an understanding of the various mechanisms of oxidation is necessary. The reactive species that can cause oxidative damage could come from one of several sources. Different oxidants will attack different sites on the protein, producing a specific set of decomposition products. Therefore, this section summarizes some of the types of oxidants known to degrade pharmaceutical proteins.

The first species to consider are organic peroxides, often found as contaminants. Organic peroxides readily react with methionine residues to produce methionine sulfoxide through a nucleophilic substitution reaction. One common source of organic peroxides is the stopper, another is the tubing used in some intravenous infusion systems. For example, silicon rubber tubing contains reactive species, such as 2,4-dichlorobenzoyl peroxide, which remains from the polymerization reaction used to manufacture the material. In addition, organic peroxides can be found in excipients, such as Tweens or polysorbates, where the peroxides are left over from the bleaching process (Nguyen, 1994).

The second reactive oxygen species to consider is singlet oxygen ($^1O_2$). Normally, molecular oxygen is in a triplet state. Upon excitation by light, the much more reactive singlet oxygen is generated. For this reason, photostability testing is becoming much more pervasive (Fransson and Hagman, 1996; Fransson, 1997; Lam et al., 1997). Singlet oxygen can potentially cause oxidation of a number of amino acid side chains, including methionine, histidine, and tryptophan (Li et al., 1995a). This type of oxidation can be prevented by removal of oxygen and/or protection of the protein from light.

A third source of reactive oxygen species in pharmaceutical dosage forms is trace metals. Redox-active metals are found as trace contaminants in additives such as buffer salts and sugars. The most common and well characterized of the redox metals and their ability to oxidize proteins are iron and copper. Both Fe(III) and Cu(II) can cause oxidation of methionine, cysteine, histidine, and tryptophan (Li et al., 1995a). Metal-catalyzed oxidation (MCO) generally occurs at metal binding sites within proteins; the reactive species are generated at or near the binding site of the metal and react quickly with labile residues within close proximity, making the identification of the exact reactive oxygen species involved difficult to identify (Li et al., 1995a, Li et al., 1995c). Thus, residues within close proximity to metal binding sites are more prone to MCO, and this process is referred to as a "caged" or site-specific process. For example, the rate of methionine oxidation in the peptide His-Met is greater than in the peptide $(Gly)_3$-Met-$(Gly)_3$, due to the fact that histidine is known to be a good ligand for Fe(III) (Li et al., 1995c).

Except for MCO reactions, oxidation processes are usually referred to as non-site-specific. Numerous examples have shown that reactivity of individual residues is greater if the residue is surface exposed compared to it being buried within the protein interior (Pearlman and Bewley, 1993; Teh et al., 1987; Liu et al., 1998; Houghten et al., 1977). For example, in human growth hormone, the reactivity of the three methionines to hydrogen peroxide correlates to their proximity to solvent accessible surfaces, and oxidation of the least solvent accessible methionine (Met-170) can usually only be seen upon unfolding of the protein in 8 M urea (Houghten et al., 1977; Nguyen, 1994). In an IgG monoclonal antibody, only one or two of the 28 tryptophans will photooxidize, and these are located on the surface and exposed to solvent (Mach et al., 1994). There are few, if any, examples of neighboring sequence effects on non-site-specific oxidation. Most likely this is due to the fact that higher order structure, including secondary, tertiary, or quaternary structure plays a much more important role than primary sequence in affecting non-site-specific oxidation rates in proteins.

Recombinant human relaxin contains two methionines, Met B(4) and Met B(25). This protein has been characterized in terms of reactivity both towards site-specific and non-site-specific oxidation, and illustrates the differences that can be seen between these two types of oxidation. In oxidation studies of human relaxin using exogenously added H_2O_2 it can be demonstrated that the more exposed Met B(25) reacts faster than the sterically hindered Met B(4) (Nguyen et al., 1993). However, when MCO of human relaxin was studied using an ascorbate/Fe(III) or Cu(II) system, the reactivity of the two methionines was reversed when compared to the hydrogen peroxide studies (Li et al., 1995b). Methionine B(4) is located near Asp and Glu residues, which can bind the metals and thus sequester the reactive species near this methionine, while B(25) is not near to any charged side chains. The His residue located at A(12) is the residue most susceptible to metal catalyzed oxidation, due to its affinity for metals despite its partially buried position (~57% surface exposed).

The example of human relaxin illustrates that the source of oxidative damage must be taken into account when considering whether changes in conformation can control oxidation rates. It is unlikely that any form of structural control could reduce MCO rates at Met B(4) or at His A(12). However, it may be possible to reduce the hydrogen peroxide generated oxidative damage to Met B(25) and especially Met B(4) if a structural stabilizer could diminish access of these residues to bulk solvent.

Effects of Oxidation of Surface and Buried Methionines on Protein Structure

It has recently been proposed that, *in vivo*, surface methionines serve to act as scavengers of oxidants in order to protect partially or fully buried residues necessary for proper folding or function of the protein (Levine et al., 1996). Ovine prolactin has seven methionine residues, of which four are oxidized readily by hydrogen peroxide, and the remaining three are only oxidized by H_2O_2 upon denaturation in 8 M urea. Oxidation of the four surface residues has little impact upon the activity or physical properties of the protein. In contrast, oxidation of remaining three reduces the bioactivity, changes the physiochemical properties, and enhances the sensitivity of the protein to cleavage by trypsin (Houghten and Li, 1976). This illustrates some of the general physiochemical consequences that may be seen upon oxidation of methionines. If a methionine residue is near the surface, then oxidation will have will generally have little effect on protein structure, and usually the only physiochemical consequence will be an increase in hydrophilicity of the protein, due to the fact methionine sulfoxide is more hydrophilic than methionine. If an oxidized methionine is partially or fully buried, then the local increase in hydrophilicity may disrupt hydrophobic interactions potentially causing loosening or unfolding of hydrophobic clusters in the protein (Vogt, 1995; Lilova et al., 1987).

Parathyroid hormone contains two methionine residues. Met-8 is buried in a hydrophobic pocket, and is less prone to oxidation than the more exposed Met-18. Oxidation of Met-18 does not alter the conformation or activity of the protein, however, oxidation of Met-8 causes a conformational change, loss in activity, and increase in hydrophobicity of the peptide, due to exposure of tryptophans previously buried within the hydrophobic pocket (Zull et al., 1990; Nabuchi et al., 1995). In the example of parathyroid hormone, it is unlikely that controlling structure would affect oxidation of the surface exposed Met-18. On the other hand, the partially buried Met-8 poses an interesting question; if oxidation of this methionine causes unfolding, then is some sort of partial unfolding or loosening of the structure necessary for it to oxidize? If this indeed were true, then oxidation of Met-8 in parathyroid hormone, or other similarly partially buried residues in other systems could possibly slowed by a general means of structural stabilization.

Limiting Solvent Accessibility of Residues

It would be of interest if a general means of structural control could be employed in order to reduce rates of oxidative degradation. Ideally, this method could be applied to a number of systems that fit the criteria of having residues prone to non-site-specific oxidation residing in a partially or fully buried position in the protein. As mentioned earlier, rates of non-site-specific oxidation correlate well with solvent accessibility. Thus, any means that could reduce the average solvent accessibility of a residue should slow its rate of oxidative degradation.

Two types of excipients could be effective at limiting solvent accessibility. The first would be a solute that is preferentially excluded from the surface of the protein, a so-called non-specific ligand or stabilizer. This type of compound and its thermodynamic effects on proteins have been described in great detail in a number of reviews (Lee and Timasheff, 1981; Timasheff, 1998; Wills et al., 1993). Examples of preferentially excluded solutes are sugars, amino acids, and polymers. Without detailing the mechanism, which is described in other chapters in this book, it is sufficient to say that addition an excluded solute to a solution will result in a protein population in which the most compact state is favored. For interferon-γ, the reduction in aggregation rate upon addition of sucrose has been directly correlated with a reduction in the population of an expanded state intermediate, which is only slightly larger size than the native state (Kendrick et al., 1998). This general phenomenon can be envisioned as the adjustment of the native state ensemble away from loose, open structures to more rigid and compact structures upon the addition of an excluded solute. Similar effects were seen in the case of recombinant interleukin-1 receptor antagonist (rIL1-ra), where sucrose was noted to induce a compaction of the native state (Kendrick et al., 1997). In both cases, there was an overall decrease in the total surface area of the protein.

One can imagine that shifting the native state ensemble would towards more compact species would also reduce the timed-average solvent accessibility of partially buried residues, lowering their susceptibility to oxidation. Such changes in the time-averaged conformation of the protein structure can be measured by monitoring the accessibility of amide hydrogens to bulk solvent via hydrogen to deuterium exchange experiments (Clarke and Itzhake, 1998). A more compact structure will have slower H-D exchange rates, directly reflecting the reduction in solvent accessibility of certain amide groups. The residues that will have their exchange rates lowered by excluded solutes represent amide protons that are partially or fully buried in the native structure; the reduction in rate may either represent protection from global unfolding or from fluctuations in the native state (Foord and Leatherbarrow, 1998). If such an effect in H-D exchange rates can be seen with addition of excluded solutes, then the next logical step is to assume that chemical reactivity of partially exposed or buried groups would likewise be reduced, as was seen in the case of cysteine reactivity in rIl1-ra (Kendrick et al., 1997). It would be simple hypothesis to test, and amenable to a number of systems. A correlation between H-D exchange rates and rates of non-site-specific oxidative degradation at partially buried residues would indeed indicate if this approach is valid.

Another aspect of the excluded solute effect is that these additives tend to favor the assembled state of multimeric systems. If a protein or peptide has a tendency to self-associate, this phenomenon would have the effect of shifting the equilibrium towards the more highly associated system. Again, this shift would result in net reduction in overall protein surface area, thereby limiting solvent accessibility of residues exposed in the on the contact sights between monomers.

The second type of ligand that could limit solvent accessibility would be a molecule (i.e., a ligand) that would bind specifically and with high affinity to the native state of the protein. Such binding would have the same thermodynamic consequence as preferential solute exclusion; the compact native conformation that most readily bound ligand would be most favored, and time-averaged solvent exposure of partially or fully buried residues would be reduced. In addition, the binding event would block a certain portion of the surface of the protein, limiting exposure to exogenous oxidants. Conversely, oxidation in the absence of the ligand would be expected to alter ligand binding to the protein. Examples of both excluded solute effects and direct ligand binding, and their effects on oxidation rates, are given below.

Conformational Control of Oxidation in Aqueous Solution

As described above, non-site-specific oxidation occurs fastest at solvent exposed residues, slowest at buried residues, and at an intermediate rate at partially exposed residues. Thus, the goal of structural control in reducing oxidation rates would be to reduce exposure of the sensitive side chains to potential oxidants. First, consider proteins that have specific endogenous ligands and the effect that ligand binding has upon oxidation.

Methionine residues in α/β-type acid-soluble spore proteins (SASP) of *Bacillus* species are readily oxidized *in vitro* by H_2O_2 or the organic peroxide, t-butyl hydroperoxide (tBHP) (Li et al., 1995a). Oxidation rendered α/β-SASP unable to bind to DNA and, presumably not confer UV and heat protection to spores *in vivo*. However, DNA binding appears to protect the protein from methionine oxidation, by rendering the methionine residues inaccessible to solvent. Therefore, this system demonstrates all of the characteristics outlined above. Ligand binding drastically reduced exposure of sensitive methionine residues. Conversely, oxidation of those residues prior to DNA binding has a profound impact on biolgical activity.

Similarly, fibronectin binds with high affinity to collagen, and within the binding region are a number of methionine residues. Oxidation of fibronectin with chloramine T greatly reduces binding affinity, although no conformational change was noted upon oxidation. However, when fibronectin is mixed with gelatin and subsequently treated with chloramine T, the methionines are effectively removed from the bulk sovlent due to binding, and no oxidation is observed (Miles and Smith, 1993). Thus, simple binding of a ligand and subsequent removal of methionines from the bulk solvent has rendered them inaccessible to oxidants.

While it is easy to appreciate that binding to a large macromolecular substrate is effective at diminishing solvent exposure, one might attribute this behavior solely to steric factors. However, binding of small ligands can also affect sensitivity to oxidation. For example, in recombinant coagulation factor VII (FVIIa), two of the protein's four methionines have been reported to undergo oxidation to methionine sulfoxide upon exposure to hydrogen peroxide (Kornfelt et al., 1999). At pH 7.5–9.5, both of these surface exposed methionines, Met-298 and Met-306, have equivalent rates of oxidation. However, upon adjustment of the pH to 5.5, the oxidation rate of Met-298 is considerably slower. Although no direct structural evidence is given, the authors conclude that, under the acidic conditions, a conformational change has occurred around the region of Met-298, partially protecting it from the solution and thus from the oxidant. In this case, the specific stabilizing ligand is hydrogen ion, even though it has no steric bulk. Here direct ligand binding to the native state causes a putative change in conformation, slowing the oxidation of Met-298.

There are few examples of excluded solute effects on oxidation. However, recent examples suggest it is, in fact, a valid approach for reducing oxidative damage. For example, DePaz et al. (2000) have reported using an excluded solute, sucrose, to inhibit methionine oxidation in the subtilisin used in laundry detergents. This protease, which has a partially buried methionine residue located near its active site, undergoes rapid oxidation at elevated pH in the presence of borate buffer. The borate is converted into a perborate by hydrogen peroxide, which carries the oxidizing species directly into the active site. Addition of sucrose slowed the oxidation rate appreciably. The authors demonstrated that the effect was not due to radical scavenging or viscosity increases. Moreover, they showed that the oxidation-competent species was slightly expanded, with a surface area

3.9% greater than the native state. Therefore, it is possible to diminish the amount of expanded, degradation-competent species in the native state ensemble and produce a more stable solution formulation, whether the pathway involves physical changes, such as aggregation, or chemical reactions, such as oxidation.

As described above, another effect of preferentially excluded ligands is to drive systems towards self-assembly, as with multimeric proteins. Assembly into a multimer should protect methionines from the solvent. In other words, control the association state of a multimeric protein can be a method for slowing non-site specific oxidation in proteins. For example, *E. coli* ribosomal protein L12 can be oxidized at its three methionine residues by hydrogen peroxide (Brot and Weissbach, 1983). This oxidation renders the protein inactive. The native, active form of L12 is a dimer, and the inactive form appears to be a monomer by ultracentrifugal analysis (Caldwell et al., 1978). Oxidation appears to have rendered the monomer incapable of forming the active dimer, and reduction of the methionine sulfoxides back to methionine by methionine sulfoxide reductase and dithiothreotol (DTT) yielded the active dimer again. Either a specific ligand or an excluded solute that could shift the equilibrium towards the dimer would slow oxidation.

Furthermore, we observed that addition of excluded solutes, like disaccharides, can affect the assembly state of the 41 amino acid peptide, human corticotropin-releasing factor, hCRF (Meyer et al., 1996). Other researchers have reported that CRF has a propensity to form tetramers at higher peptide concentrations, as well as form larger multimers under certain conditions (Dathe et al., 1996). Although specific details of the oligomerization need to be determined, it appears that solution conditions can be adjusted to push the system to a more highly associated state (unpublished results). The two methionines in the peptide, Met-21 and Met-38, have been found to oxidize upon storage in a liquid formulation at different rates. Met-38, located near the C-terminus, oxidizes several times faster than Met-21 at a peptide concentration of 1 mg/mL. Upon increasing the concentration to 8 mg/mL, the rate of oxidation of Met-38 is slowed almost four-fold, while the Met-21 rate is slowed only by a factor of ~1.4. We suggest that both methionines become progressively more protected from the solvent as the peptide associates, with the extent of solvent exposure changing the most for Met-38, as it becomes more deeply buried under solution conditions which push the peptide toward a relatively large oligomer.

Structural Control of Oxidation in Lyophilized Products

Due to the fact that many proteins do not possess adequate stability in solution in order to provide a reasonable shelf life, many protein pharmaceuticals are prepared as lyophilized products. Although lyophilization may solve some of the physical instability problems, proteins still may also undergo chemical degradation in the solid state. Certainly, chemical degradation is typically slower in the solid state than in liquid state (Li et al., 1995a; Kenkare and Richards, 1966; Chang et al., 1996), due to reduced mobility of the protein and limited exposure to water. However, this is not always the case (Fransson et al., 1996). Regardless of whether oxidation in retarded in a lyophilized cake compared to a solution, oxidation remains a major degradative pathway that must be considered. Variables such as residual moisture, type of excipients, and composition of the headspace gas in relation to oxidation have previously been discussed in a number of research and review articles (Li et al., 1995a; Pikal et al., 1991; Fransson et al., 1996). As in liquid formulations, the effect of structure on rates of oxidation in the solid state should not be ignored. Maintenance of native protein structure in the dried state may serve to reduce chemical instability, as well as physical instability, assuming that surface exposure would have similar effects on oxidation as solvent exposure does in the liquid state. Therefore, proper design of a freeze dried formulation in which the maintenance of native secondary structure to reduce physical instability problems such as aggregation may have the secondary benefit of reducing surface exposure of labile side chains in the solid form as well.

For example, in hGH, irreversible changes during freeze-drying were minimal, but upon storage, aggregation and chemical decomposition (via methionine oxidation and asparagine damidation became problematic) (Pikal et al., 1991). The formulation that contained no excipients had the highest level of aggregates, as well as the highest level of methionine sulfoxide. The final formulation, which conferred the greatest stability, was a combination of mannitol and glycine, where the glycine remained amorphous. This formulation not only provided stabilization against aggregation, but against methionine oxidation as well. At the very least, this correlation suggests that unfolding of the protein in the dried state, which leads to increased susceptibility to aggregation, also may also make the protein more susceptible to oxidation. Whether or not this correlation between maintenance of native-like structure in the solid state and lower rates of oxidation holds true in all cases remains to be seen. Such reports in the open literature are rare. In addition, other variables, such as the diffusivity of reactants in the solid, the storage temperature relative to the glass transition temperature, and the exact nature of reactive species in the dried state, also need to be considered as well.

Summary of Structural Control of Oxidation

Oxidation of the various side chains of proteins and peptides remains a major problem to overcome in formulation science. A rational approach stemming from an understanding of oxidative mechanisms and oxidative sources is already fairly well understood, and has helped in a number of cases where oxidative degradation was occurring in a pharmaceutical dosage form. Consideration and control of secondary, tertiary and quaternary structure will hopefully be yet an additional method to slow oxidation and increase shelf life. In the case of liquid formulations, this could be done by adjustment of specific solution conditions such as pH in order to alter conformational states, reduction of solvent exposure of methionines via association states, or addition of excluded solutes to produce a more compact native ensemble. The work by DePaz et al. (2000) demonstrates that for buried residues the oxidation-competent species is slightly larger than the native state. The use of excluded solutes to shift the equilibrium towards to more compact species should and does slow chemical degradation. We expect other examples to emerge in the near future.

In lyophilized formulations structural control may already be in place in many instances, as maintenance of a native-like secondary structure is already a requirement for reducing aggregation from physical instability, and its effect should certainly be considered if oxidation is a route of degradation. It has been shown that less mobility (i.e., in an amorphous glass below the glass transition temperature) does slow deamidation (Chang et al., 1996).

SUMMARY

Control of chemical instability in protein pharmaceuticals continues to be a critical issue in developing stable formulations. While the effects of pH, buffer composition, ionic strength and temperature remain the most effective methods for controlling hydrolysis and oxidation reactions, it appears that conformational control may also be important. Addition of excipients to maintain native structure and reduce the propensity of the protein to denature and/or aggregate is already a central theme in stabilizing proteins (Arakawa et al., 1993). The same additives have now been found to slow both deamidation and oxidation, whether in solution or in the solid state. What is emerging is an additional approach for producing protein pharmaceuticals that maintain native structure and activity during long-term storage.

REFERENCES

Arakawa, T., Prestrelski, S.J., Kenney, W.C., and Carpenter, J.F., 1993. Factors affecting short-term and long-term stabilites of proteins, *Adv. Drug Delivery Rev.* 10: 1–28.

Araki, F., Nakamura, H., Nojima, N., Tsukumo, K., and Sakamoto, S., 1989. Stability of recombinant human epidermal growth factor in various solutions, *Chem. Pharm. Bull.* 37:404–406.

Artigues, A., Birkett, A., and Schirch, V., 1990. Evidence for the in vivo deamidation and isomerization of an asparaginyl residue in cytosolic serine hydroxymethyltransferase, *J. Biol. Chem.* 265:4853–4858.

Aswad, D.W., ed., 1995. *Deamidation and Isoasparate Formation in Peptides and Proteins*, CRC Press, Boca Raton.

Becker, G.W., Tackitt, P.M., Bromer, W.W., Lefeber, D.S., and Riggin, R.M., 1988. Isolation and characterization of a sulfoxide and desamido derivative of biosynthetic human growth hormone, *Biotechnol. Appl. Biochem.* 10:326–337.

Berson, S.A. and Yalow, R.S., 1966. Deamidation of insulin during storage in frozen state, *Diabetes* 15:875–878.

Bongers, J., Heimer, E.P., Lambros, T., Pan, Y.C., Campbell, R.M., and Felix, A.M., 1992. Degradation of aspartic acid and asparagine residues in human growth hormone-releasing factor, *Int. J. Peptide Protein Res.* 39:364–374.

Brange, J., Langkj¾ r, L., Havelund, S., and V¿lund, A., 1992. Chemical stability of insulin. 1. hydrolytic degradation during storage of pharmaceutical preparations. *Pharm. Res.* 9:715–726.

Brennan, T.V. and Clarke, S., 1995. Effect of adjacent histidine and cysteine residues on the spontaneous degradation of asparaginyl- and aspartyl-containing peptides, *Int. J. Peptide Protein Res.* 45:547–553.

Brot, N. and Weissbach, H., 1983. Biochemistry and physiological role of methionine sulfoxide residues in proteins, *Arch. Biochem. Biophys.* 223:271–281.

Caldwell, P., Luk, D.C., Weissbach, H., and Brot, N., 1978. Oxidation of the methionine residues of Escherichia coli ribosomal protein L12 decreases the protein's biological activity, *Proc. Natl. Acad. Sci. USA* 75:5349–5352.

Chang, B.S., Beauvais, R.M., Dong, A., and Carpenter, J.F., 1996. Physical factors affecting the storage stability of freeze-dried interleukin-1 receptor antagonist: glass transition and protein conformation, *Arch. Biochem. Biophys.* 331:249–258.

Chen, B., Costantino, H.R., Hsu, C.-C., and Shire, S.J., 1999. Influence of calcium ions on the structure and stability of recombinant human deoxyribonuclease I in the aqueous and lyophilized States, *J. Pharm. Sci.* 88:477–482.

Chloupek, R.C., Harris, R.J., Leonard, C.K., Keck, R.G., Keyt, B.A., Spellman, M.W., Jones, A.J.S., and Hancock, W.S., 1989. Study of the primary structure of recombinant tissue plasminogen activator by reversed phase high performance liquid chromatographic tryptic mapping, *J. Chromatogr.* 463:375–396.

Clarke, S., 1987. Propensity for spontaneous succinimide formation from aspartyl and asparaginyl residues in cellular proteins, *Int. J. Peptide Protein Res.* 30:808–829.

Clarke, J. and Itzhake, L.S., 1998. Hydrogen exchange and protein folding, *Curr. Opinion Struct. Biol.* 8:112–118.

Cleland, J.L., Mac, A., Boyd, B., Yang, J., Duenas, E.T., Yeung, D., Brooks, D., Hsu, C., Chu, H., Mukku, V., and Jones, A.J.S., 1997. The stability of recombinant human growth hormone in poly(lactic-co-glycolic acid) (PLGA) microspheres, *Pharm. Res.* 14:420–425.

Cleland, J.L., Powell, M.F., and Shire, S.J., 1993. The development of stable protein formulations: a close look at protein aggregation, deamidation, and oxidation, *Crit. Rev. Ther. Drug Carrier Systems* 10:307–377.

Darrington, R.T. and Anderson, B.D., 1994. The role of intramolecular nucleophilic catalysis and the effects of self-association on the deamidation of human insulin at low pH, *Pharm. Res.* 11:784–793.

Dathe, M., Fabian, H., Gast, K., Zirwer, D., Winter, R., Beyermann, M., Schumann, M., and Bienert, M., 1996. Conformational differences of ovine and human corticotropin releasing hormone. A CD, IR, NMR and dynamic light scattering study, *Int. J. Peptide Protein Res.* 47:383–393.

DePaz, R.A., Barnett, C.C., Carpenter, J.F., Dale, D.A., Gaertner, A., and Randolph, T.W., 2000. The excluding effects of sucrose on a protein chemical degradation pathway: methionine oxidation in subtilisin, *Arch. Biochem. Biophys.* 384:123–132.

DiAugustine, R.P., Gibson, B.W., Aberth, W., Kelly, M., Ferrua, C.M., Tomooka, Y., Brown, C.F., and Walker, M., 1987. Evidence for isoaspartyl (deamidated) forms of mouse epidermal growth factor, *Anal. Biochem.* 165:420–429.

Di Donato, A., Galletti, P., and D'Alession, G., 1986. Selective deamidation and enzymatic methylation of seminal ribonuclease, *Biochemistry* 25:8361–8368.

Fisher, B.D. and Porter, P.B., 1981. Stability of bovine insulin, *J. Pharm. Pharmacol.* 33:203–206.

Foord, R.L. and Leatherbarrow, R.J., 1998. Effect of osmolytes on the exchange rates of backbone amide protons in proteins, *Biochemistry* 37:2969–2978.

Fransson, J., Florin-Robertsson, E., Axelsson, K., and Nyhlen, C., 1996. Oxidation of human insulin-like growth factor I in formulation studies: kinetics of methionine oxidation in aqueous solution and in solid state, *Pharm. Res.* 13:1252–1257.

Fransson, J. and Hagman, A., 1996. Oxidation of human insulin-like growth factor I in formulation studies, II. Effects of oxygen, visible light, and phosphate on methionine oxidation in aqueous solution and evaluation of possible mechanisms, *Pharm. Res.* 13:1476–1481.

Fransson, J.R., 1997. Oxidation of human insulin-like growth factor I in formulation studies. 3. Factorial experiments of the effects of ferric ions, EDTA, and visible light on methionine oxidation and covalent aggregation in aqueous solution, *J. Pharm. Sci.* 86:1046–1050.

Frenz, J., Wu, S.-L., and Hancock, W.S., 1989. Characterization of human growth hormone by capillary electrophoresis, *J. Chromatogr.* 480:379–391.

Geiger, T. and Clarke, S., 1987. Deamidation, isomerization, and racemization at asparaginyl and aspartyl residues in peptides. Succinimide-linked reactions that contribute to protein degradation, *J. Biol. Chem.* 262:785–794.

George-Nascimento, C., Lowenson, J., Borissenko, M., Calderon, M., Medina-Selby, A., Kuo, J., Clarke, S., and Randolph A., 1990. Replacement of a labile aspartyl residue increases the stability of human epidermal growth factor, *Biochemistry* 29:9584–9591.

Gietz, U., Alder, R., Langguth, P., Arvinte, T., and Merkle, H.P., 1998. Chemical degradation kinetics of recombinant hirudin (HV1) in aqueous solution: effect of pH, *Pharm. Res.* 15:1456–1462.

Hayes, C.S., Illades-Aguiar, B., Casillas-Martinez, L., and Setlow, P., 1998. In vitro and in vivo oxidation of methionine residues in small, acid-soluble spore proteins from Bacillus species, *J. Bacteriol.* 180:2694–2700.

Houghten, R.A. and Li, C.H., 1976. Studies on pituitary prolactin. 39. reaction of the ovine hormone with hydrogen peroxide, *Biochim. Biophys. Acta* 439:240–249.

Houghten, R.A., Glaser, C.B., and Li, C.H., 1977. Human somatotropin. Reaction with hydrogen peroxide. Arch. Biochem. Biophys. 178:350–355.

Hsu, Y.R., Chang, W.C., Mendiaz, E.A., Hara, S., Chow, D.T., Mann, D.B., Langley, K.E., and Lu, H.S., 1998. Selective deamidation of recombinant human stem cell factor during in vitro Aging: isolation and characterization of the aspartyl and isoaspartyl homodimers and heterodimers, *Biochemistry* 37:2251–2262.

Johnson, B.A., Shirokawa, J.M., Hancock, W.S., Spellman, M.W., Basa, L.J., and Aswad, D.W., 1989. Formation of isoaspartate at two distinct sites during in vitro aging of human growth hormone, *J. Biol. Chem.* 264:14262–14271.

Kendrick, B.S., Chang, B.S., Arakawa, T., Peterson, B., Randolph, T.W., Manning, M.C., and Carpenter, J.F., 1997. Preferential exclusion of sucrose from recombinant interleukin-1 receptor antagonist: Role in restricted conformational mobility and compaction of native state. *Proc. Natl. Acad. Sci. USA* 94:11917–11922.

Kendrick, B.S., Carpenter, J.F., Cleland, J.L., and Randolph, T.W., 1998. A transient expansion of the native state preces aggregation of recombinant human interferon-γ. *Proc. Natl. Acad. Sci. USA* 95:14142–14146.

Kenkare, U.W. and Richards, F.M., 1966. The histidyl residues in ribonuclease-S. Photooxidation in solution and in single crystals, the iodination of histidine-12, *J. Biol. Chem.* 241:3197–3206.

Kornfelt, T., Persson, E., and Palm, L., 1999. Oxidation of methionine residues in coagulation factor VIIa, *Arch. Biochem. Biophys.* 363:43–54.

Kosen, P.A., 1992. Disulfide bonds in proteins, in: *Stability of Protein Pharmaceuticals, Part A: Chemical and Physical Pathways of Protein Degradation*, T.J. Ahern and M.C. Manning, eds., Plenum Press, New York, pp. 31–66.

Kossiakoff, A.A., 1988. Tertiary structure is a principal determinant to protein deamidation, *Science* 240:191–194.

Kroon, D.J., Baldwin-Ferro, A., and Lalan, P., 1992. Identification of sites of degradation in a therapeutic monoclonal antibody by peptide mapping, *Pharm. Res.* 9:1386–1393.

Lam, X.M., Yang, J.Y., and Cleland, J.L., 1997. Antioxidants for prevention of methionine oxidation in recombinant monoclonal antibody HER2, *J. Pharm. Sci.* 86:1250–1255.

Lee, J.C. and Timasheff, S.N., 1981. The stabilization of proteins by sucrose, *J. Biol. Chem.* 256:7193–7201.

Levine, R.L., et al., 1996. Methionine residues as endogenous antioxidants in proteins, *Proc. Natl. Acad. Sci. USA* 93:15036–15040.

Lewis, U.J., Cheever, E.V., Hopkins, W.C., 1970. Kinetic study of the deamidation of growth hormone and prolactin, *Biochim. Biophys. Acta.* 214:498–508.

Li, S., Schoneich, C., and Borchardt, R.T., 1995a. Chemical instability of protein pharmaceuticals: mechanisms of oxidation and strategies for stabilization. *Biotechnol. Bioeng.* 48:490–500.

Li, S., Nguyen, T.H., Schoneich, C., and Borchardt, R.T., 1995b. Aggregation and precipitation of human relaxin induced by metal-catalyzed oxidation, *Biochemistry* 34:5762–5772.

Li, S., Schoneich, C., and Borchardt, R.T., 1995c. Chemical pathways of peptide degradation. VIII. Oxidation of methionine in small model peptides by prooxidant/transition metal ion systems: influence of selective scavengers for reactive oxygen intermediates. *Pharm. Res.* 12:348–355.

Lilova, A., Kleinschmidt, T., and Nedkov, P., 1987. Methionine residue susceptibility in native subtilisin DY, *Biol. Chem. HoppeSeyler* 368:513–519.

Liu, D.T.-Y., 1992. Deamidation: a source of microheterogeneity in pharmaceutical proteins, *Trends Biotechnol.* 10:364–369.

Liu, J.L., Lu, K.V., Eris, T., Katta, V., Westcott, K.R., Narhi, L.O., and Lu, H.S., 1998. In vitro methionine oxidation of recombinant human leptin, *Pharm. Res.* 15:632–640.

Lura, R., Schirch, V., 1988. Role of peptide conformation in the rate and mechanism of deamidation of asparaginyl residues, *Biochemistry* 27:7671–7677.

Mach, H., et al., 1994. Origin of the photosensitivity of a monoclonal immunoglobulin G, in: *Formulation and Delivery of Proteins and Peptides*, J.L. Cleland and R. Langer, eds., American Chemical Society Symp. Series, Washington, D.C., pp. 72–84.

Manning, M.C., Patel, K., Borchardt, R.T., 1989. Stability of protein pharmaceuticals, *Pharm. Res.* 6:903–918.

McCrossin, L.E., Charman, W.N., and Charman, S.A., 1998. Degradation of recombinant porcine growth hormone in the presence of guanidine hydrochloride, *Int. J. Pharm.* 173:157–170.

Meyer, J.D., Ho, B., Snyder, S., and Manning, M.C., 1996. Solution behavior of human corticotropin releasing factor, *Pharm. Res.* 13:S-100.

Miles, A.M. and Smith, R.L., 1993. Functional methionines in the collagen/gelatin binding domain of plasma fibronectin: effects of chemical modification by chloramine T, *Biochemistry* 32:8168–8178.

Nabuchi, Y., Fujiwara, E., Ueno, K., Kuboniwa ,H., Asoh, Y., and Ushio, H., 1995. Oxidation of recombinant human parathyroid hormone: effect of oxidized position on the biological activity, *Pharm. Res.* 12:2049–2052.

Nguyen, T.H., Oxidative degradation of protein pharmacetucals, in: *Formulation and Delivery of Proteins and Peptides*, J.L. Cleland and R. Langer, eds., American Chemical Society Symp. Series, Washington, D.C., pp. 59–71.

Nguyen, T.H., Burnier, J., and Meng, W., 1993. The kinetics of relaxin oxidation by hydrogen peroxide. *Pharm. Res.* 10:1563–1571.

Nielsen, R.G., Sittampalam, G.S., and Rickard, E.C., 1989. Capillary zone electrophoresis of insulin and growth hormone, *Anal. Biochem.* 177:20–26.

Oliva, A., Farina, J.B., and LlabrŽs, M., 1996. Influence of temperature and shaking on stability of insulin preparations: degradation kinetics, *Int. J. Pharm.* 143:163–170.

Patel, K. and Borchardt, R.T., 1990. Chemical pathways of peptide degradation. II. kinetics of deamidation of an asparaginyl residue in a model hexapeptide, *Pharm. Res.* 7:703–711.

Pearlman, R. and Bewley, T.A., 1993. Stability and characterization of human growth hormone, in: *Pharmaceutical Biotechnology: Case Studies*, Vol. 5, Y.J. Wang and R. Pearlman, eds., Plenum Press, New York, New York. pp. 1–59.

Pearlman, R. and Nguyen, T.H., 1992. Pharmaceutics of protein drugs, *J. Pharm. Pharmacol.* 44 (Suppl 1):178–185.

Pikal, M.J., Dellerman, K.M., Roy, M.L., and Riggin, R.M., 1991. The effects of formulation variables on the stability of freeze-dried human growth hormone, *Pharm. Res.* 8:427–436.

Senderoff, R.I., Wootton, S.C., Boctor, A.M., Chen, T.M., Giordani, A.B., Julian, T.N., and Radebaugh, G.W., 1994. Aqueous stability of human epidermal growth factor 1–48, *Pharm. Res.* 11:1712–1720.

Sharma, S., Hammen, P.K., Anderson, J.W., Leung, A., Georges, F., Hengstenberg, W., Klevit, R.E., and Waygood, E.B., 1993. Deamidation of HPr, a phosphocarrier protein of the phosphoenolpyruvate:sugar phosphotransferase system, involves asparagine 8 (HPr-1) and asparagine 12 (HPr-2) in isoaspartyl acid formation, *J. Biol. Chem.* 268:17695–17704.

Son, K. and Kwon, C., 1995. Stabilization of human epidermal growth factor (hEGF) in aqueous solution, *Pharm. Res.* 12:451–454.

Stevenson, C.L., Donlan, M.E., Friedman, A.R., and Borchardt, R.T., 1993a. Solution conformation of Leu27-hGRF(1–32)-NH_2 and Its deamidation products by 2D NMR, *Int. J. Peptide Protein Res.* 42:24–32.

Stevenson, C.L., Friedman, A.R., Kubiak, T.M., Donlan, M.E., and Borchardt, R.T., 1993b. Effect of secondary structure on the rate of deamidation of several growth hormone releasing factor analogs, *Int. J. Peptide Protein Res.* 42:497–503.

Strickley, R.G. and Anderson, B.D., 1996. Solid-state stability of human insulin. I. mechanism and the effect of water on the kinetics of degradation in lyophiles from pH 2–5 solutions, *Pharm. Res.* 13:1142–1153.

Teh, L.C., Murphy, L.J., Huq, N.L., Surus, A.S., Friesen, H.G., Lazarus, L., and Chapman, G.E., 1987. Methionine oxidation in human growth hormone and human chorionic somatomammotropin. Effects on receptor binding and biological activities, *J. Biol. Chem.* 262:6472–6477.

Timasheff, S.N., 1998. Control of protein stability and reactions by weakly interacting cosolvents: The simplicity of the complicated. *Adv. Protein Chem.* 91:355–432.

Tsai, P.K., Bruner, M.W., Irwin, J.I., Yu Ip, C.C., Oliver, C.N., Nelson, R.W., Volkin, D.B., and Middaugh, C.R., 1993. Origin of the isoelectric heterogeneity of monoclonal immunoglobulin h1B4, *Pharm. Res.* 10:1580–1586.

Tyler-Cross, R. and Schirch, V., 1991. Effects of amino acid sequence, buffers, and ionic strength on the rate and mechanism of deamidation of asparagine residues in small peptides, *J. Biol. Chem.* 266:22549–22556.

Violand, B.N., Schlittler, M.R., Toren, P.C., and Siegel, N.R., 1990. Formation of isoaspartate 99 in bovine and porcine somatotropins, *J. Prot. Chem.* 9:109–117.

Violand, B.N., Schlittler, M.R., Kolodziej, E.W., Toren, P.C., Cabonce, M.A., Siegel, N.R., Duffin, K.L., Zobel, J.F., Smith, C.E., and Tou, J.S., 1992. Isolation and characterization of porcine somatotropin containing a succinimide residue in place of aspartate129, *Protein Sci.* 1:1634–1641.

Vogt, W., 1995. Oxidation of methionyl residues in proteins: tools, targets, and reversal, *Free Radical Biol. Med.* 18:93–105.

Wearne, S.J. and Creighton, T.E., 1989. Effect of protein conformation on rate of deamidation: ribonuclease A, *Proteins* 5:8–12.

Wills, P.R., Comper, W.D., and Winzor, D.J., 1993. Thermodynamic nonideality in macromolecular solutions: interpretation of virial coefficients, *Arch. Biochem. Biophys.* 300:206–213.

Wright, H.T., 1991. Sequence and structure determinants of the nonenzymatic deamidation of asparagine and glutamine residues in proteins, *Protein Eng.* 4:283–294.

Xie, M. and Schowen, R.L., 1999. Secondary structure and protein deamidation, *J. Pharm. Sci.* 88:8–13.

Zull, J.E., Smith, S.K., and Wiltshire, R., 1990. Effect of methionine oxidation and deletion of amine-terminal resiudes on the conformation of parathyroid hormone, *J. Biol. Chem.* 265:5671–5676.

5

Rational Design of Stable Lyophilized Protein Formulations: Theory and Practice

John F. Carpenter[1,2], *Beyong S. Chang*[3], *William Garzon-Rodriguez*[1,2], *and Theodore W. Randolph*[1,4]

INTRODUCTION

For ease of preparation and cost containment by the manufacturer, and ease of handling by the end user, an aqueous therapeutic protein formulation usually is preferred. However, with many proteins it is not possible—especially considering the time constraints for product development—to develop sufficiently stable aqueous formulations. Unacceptable denaturation and aggregation can be induced readily by the numerous stresses to which a protein in aqueous solution is sensitive; e.g., heating, agitation, freezing, pH changes, and exposure to interfaces or denaturants (Arakawa et al., 1993; Cleland et al., 1993; Brange, 2000; Bummer

---

John F. Carpenter, William Garzon-Rodriguez, and Theodore W. Randdph • Center for Pharmaceutical Biotechnology. *John F. Carpenter, William Garzon-Rodriguez, and Theodore W. Randolph* • Department of Pharmaceutical Sciences, School of Pharmacy, University of Colorado Health Sciences Center, Denver, CO 80262. *Beyong S. Chang* • Amgen, Inc., Thousand Oaks, CA 91320. *Theodore W. Randolph* • Department of Chemical Engineering, University of Colorado, Boulder, CO 80309.

Rational Design of Stable Protein Formulations, edited by Carpenter and Manning. Kluwer Academic / Plenum Publishers, New York, 2002.

and Koppenol, 2000). Furthermore, even under conditions that thermodynamically greatly favor the native state of proteins, aggregation can arise during months of storage in aqueous solution (e.g., Gu et al., 1991; Arakawa et al., 1993; Chen et al., 1994; Volkin and Middaugh, 1996; Chang et al., 1996a). In addition, several chemical degradation pathways (e.g., hydrolysis and deamidation) are mediated by water. In aqueous formulations, the rates of these and other (e.g., oxidation) chemical degradation reactions can be unacceptably rapid on the time scale of storage (e.g., 18–24 months) for pharmaceutical products (Manning et al., 1989; Cleland et al., 1993; Goolcharran et al., 2000; Bummer and Koppenol, 2000).

In contrast, a properly lyophilized formulation can maintain adequate physical and chemical stability of the protein during shipping and long-term storage, even at ambient temperatures. As will be outlined in this chapter, developing stable lyophilized protein formulations should be a rational, straightforward process, which for most proteins should be rapid. With liquid formulation development, it may only be possible to obtain adequate protein stability after lengthy studies. Furthermore, sometimes there are conflicting conditions (e.g., pH) needed to slow sufficiently multiple degradation pathways in aqueous solution. Considering these issues plus the fact that formulation scientists now have to deal with numerous proteins and/or variants of a given protein, lyophilization should be considered as a primary mode for product development. Only if a parallel effort to develop an aqueous formulation is successful, will a final lyophilized product not be needed.

Rapid formulation development has important financial ramifications. A drug product has a finite patent life, during which time the company has an exclusive market. Considering that even a moderately successful drug product has annual sales of hundreds of millions of dollars, potentially millions of dollars in sales are lost for each day of delay in bringing a product to market. Unfortunately, there are often delays because the formulations designed during early stage development and clinical trials (e.g., frozen) were not adequate for the final product. With a rational approach to formulation development, pharmaceutical scientists and process engineers can minimize the risk of this problem and the time needed to obtain a successful, final formulation. The success of such efforts depends on frank, open communication between the groups involved. For example, it is critical that the formulation scientists learn from the process engineers the issues for large-scale lyophilization runs, which are usually conducted in units that do not have the capacity to match processing parameters obtained in a small-scale research lyophilizers.

Despite the best efforts of the scientists and engineers, all too often delays in formulation development arise because sufficient resources are not invested in product development. For example, sometimes, purchase of essential equipment (e.g., a differential scanning calorimeter), which costs a minute fraction of a day's sale of product, is not allowed. To avoid unnecessary delays in product launch, which can have disastrous consequences for the company and for patients, it is essential that companies appreciate that product development is ultimately a key limiting factor in getting a therapeutic to market. Hence, development efforts need to be as well funded as the usually much more visible drug discovery research programs. If a product is not stable, it will not be marketed, no matter how dramatic an impact it can have on human health and the financial status of the company.

MINIMAL CRITERIA FOR A SUCCESSFUL LYOPHILIZED FORMULATION

Research over the past several years has demonstrated that five criteria define the minimal conditions necessary for a successful lyophilized protein formulation (Table 1).

The first four criteria can be met with use of the appropriate excipients and lyophilization cycle design. For information on the proper design of lyophilization cycles, the reader is directed to the numerous previous reviews in this area (Franks, 1990; Pikal, 1990; Nail and Gatlin, 1993; Gatlin and Nail, 1994; Carpenter and Chang, 1996; Rey and May, 1999; Cappola, 2000). For the current chapter we will only consider cycle design in terms of the interplay between formulation physical properties (e.g., collapse temperature) and process parameters (see below). The last criterion listed in Table 1 requires insight into the unique physicochemical properties of each therapeutic protein, which will be explained in more detail below. We will discuss in turn why each of these criteria is important. Then we will present an explanation of how to design rationally a formulation to meet these criteria.

Table 1.
Minimal Criteria for a Successful Lyophilized Protein Formulation

1. Protein unfolding during freezing and drying is inhibited.
2. The glass transition temperature of the product exceeds the planned storage temperature (e.g., $T_g > 30°C$).
3. The water content is relatively low (e.g., 1% by mass).
4. A strong, elegant cake structure is obtained (i.e., collapse and meltback are avoided).
5. Steps are taken to minimize specific routes of protein chemical degradation (e.g., product vials are sealed under nitrogen to reduce the rate of methionine oxidation).

Inhibition of Lyophilization-Induced Protein Unfolding

The stresses of freezing and drying cause protein unfolding, and the formulation must be designed to inhibit unfolding at each step (Prestrelski et al., 1993a,b; Carpenter et al., 1993; Prestrelski et al., 1995; Constantino et al., 1995, 1998; Griebenow and Klibanov, 1995; Allison et al., 1996, 1998, 1999, 2000; Chang et al., 1996b; Krielgaard et al., 1998a, 1999; Chen et al., 1999; Bell, 1999; Carrasquillo et al., 2000). Even if the formulation excipients and/or intrinsic thermodynamic stability of the protein prevent denaturation during freezing, unfolding can arise during subsequent drying (Carpenter et al., 1993; Prestrelski et al., 1993b; Allison et al., 1998; Carrasquillo et al., 2000). Conversely, once a protein unfolds during freezing, it will not regain native structure during dehydration.

For many proteins, unfolding during lyophilization leads to clinically unacceptable, non-native aggregates, even when samples are rehydrated immediately after lyophilization (Prestrelski et al., 1993a, 1995; Allison et al., 1996; Krielgaard et al., 1998a, 1999; Costantino et al., 1998). Aggregates are not necessarily formed during freezing and drying. Rather, during rehydration refolding of structurally perturbed protein molecules competes with formation of non-native protein aggregates (Prestrelski et al., 1993a). Aggregation can be minimized by including stabilizing excipients (e.g., sucrose or trehalose) in the formulation to inhibit lyophilization-induced unfolding (Prestrelski et al., 1993a, 1995; Allison et al., 1996; Krielgaard et al., 1998a, 1999; Costantino et al., 1998). Furthermore, fostering refolding during rehydration (e.g., with surfactants) can reduce aggregation (Chang et al., 1996c; Zhang et al., 1995, 1996).

In addition to minimizing protein aggregation during lyophilization/rehydration, maximizing retention of native protein structure in the dried solid is essential for optimizing long-term storage stability (Prestrelski et al., 1995; Chang et al., 1996b; Krielgaard et al., 1998a, 1999; Allison et al., 2000; Cleland et al., 2001). Both chemical and physical degradation in the dried solid can be accelerated if protein unfolding is not inhibited during lyophilization. With chemical degradation, a non-native structure may provide an environment conducive to covalent modification of one or more residues. For example, exposure of a methionine, which is normally buried deep in the interior of the native protein, on the surface of an unfolded dried protein may foster oxidation. Increased levels of aggregates noted after storage and rehydration of unfolded proteins could be due to formation of non-native intermolecular contacts within the dried solid, perturbation of refolding during rehydration because of chemical degradation, and/or other undefined processes.

Infrared spectroscopy has been used routinely to compare the secondary structures of a protein in lyophilized formulations to that of the native protein in aqueous solution (Prestrelski et al., 1993a,b; Prestrelski et al., 1995; Dong et al.,

1995; Constantino et al., 1995, 1998; Griebenow and Klibanov, 1995; Allison et al., 1996, 1998, 1999, 2000; Chang et al., 1996b; Krielgaard et al., 1998a, 1999; Carpenter et al., 1998; Chen et al., 1999; Carrasquillo et al., 2000). This method should be considered essential in the development of stable lyophilized formulations, because it allows one to assess rapidly the effectiveness of formulations at inhibiting protein unfolding. Technical details about how to employ infrared spectroscopy to design stable lyophilized protein formulations can be found in the papers cited above.

Storage at Temperatures Below Formulation Glass Transition Temperature

In the dried powder, the protein is a component of an amorphous phase that includes amorphous excipients and water. If this glassy matrix is held below its characteristic glass transition temperature (T_g), the rate of diffusion-controlled reactions, including protein unfolding/aggregation and chemical degradation, are greatly reduced, relative to rates noted at temperatures $>T_g$ (Roy et al., 1991; Franks, 1990; Franks et al., 1991; Pikal, 1994, 1999). T_g can be determined with differential scanning calorimetry (DSC) or other thermal scanning methods (Nail and Gatlin, 1993; Chang and Randall, 1992; Craig and Royall, 1998; Verdonck et al., 1999).

Obtaining a formulation T_g in excess of the planned storage temperature (e.g., room temperature) is absolutely essential for optimal protein stability (e.g., Franks et al., 1991; Pikal, 1994, 1999; Carpenter and Chang, 1996; Duddu and Dal Monte, 1997). The T_g of a given amorphous phase is dependent on the T_g and mass percent of each component, including water (Angell, 1995; Franks et al., 1991; Levine and Slade, 1988, 1992; Pikal, 1994, 1999). Compared to excipients, dried proteins have relatively high T_g's (e.g., >150°C; Angell, 1995). Thus, with all other factors being held constant, the formulation T_g varies directly with the mass fraction of protein. However, care must be taken that the mass fraction of protein is not so high that there are not adequate levels of stabilizing excipients to prevent protein unfolding during lyophilization (Cleland et al., 2001; and see below).

Fortunately, sucrose and trehalose, which are the preferred excipients for inhibiting lyophilization-induced protein unfolding (see below), also provide a glassy matrix with acceptably high T_g values. For example, with water contents of 1% the T_g for pure sucrose and trehalose are about 100 and 65°C, respectively (Crowe et al., 1998).

It has now been documented with several proteins, that simply storing the formulation at temperature below T_g alone does not assure optimal stability. A native protein structure is also required. For example, proteins lyophilized in dextran alone are usually unfolded, but in a glassy matrix with a relatively high T_g (e.g., > 75° C). Yet they still degrade at relatively rapid rates compared to those for native protein molecules lyophilized with either sucrose or trehalose (Krielgaard et al., 1998a, 1999; Lueckel et al., 1998; Allison et al., 2000; Yoshioka et al., 2000). On pharmaceutical time scales of several months of storage many degradative reactions are not coupled to the glass transition of a formulation. This is because on these times scales there is still significant molecular mobility, even at temperatures well below (e.g., more than 30° C) the T_g (Hancock et al., 1995; Duddu et al., 1997; Pikal, 1999; Yoshioka et al., 1999).

The Water Content is Relatively Low

Because of its very low T_g (−135°C), water is a potent plasticizer for glasses; increasing water content in the dried formulation will greatly reduce T_g. For example, increasing the water content of pure sucrose from 1 to about 3–4% (g H_20/100 g dried powder) is sufficient to reduce the T_g to below room temperature (Crowe et al., 1998). It is critical to achieve a sufficiently low water level for a given formulation such that T_g exceeds the planned storage temperature. The lyophilization cycle dictates the initial water content (see reviews listed above). The most important parameter is the temperature for secondary drying, when the unfrozen water is desorbed (Pikal et al., 1990).

Water can also be transferred to the product from the vial stoppers during storage (Pikal and Shah, 1992; DeGrazio and Flynn, 1992; Hora and Wolfe, 1999). This effect can be dramatic. For example, let's consider a formulation containing 10 mg of dried protein, 40 mg of sucrose and initial water content of 1% by weight. The total amorphous fraction containing protein and sucrose has 0.50 mg of water. If 1.0 mg of water was transferred from the stopper to the product, the water content of this fraction would increase from 1% to 3.0%. This increase would be sufficient to lower the formulation T_g to below room temperature (Crowe et al., 1998). The risk of transfer of moisture from stoppers can be minimized by drying the stoppers before use, and, if acceptable for a given product, using stoppers coated with a material such as Teflon (see Hora and Wolfe, 1999).

A Strong, Elegant Cake Structure is Obtained

Often the most desired cake has strong, porous structure, without macroscopic collapse or meltback. This structure has a high surface area to volume ratio, which aids in the rapid dissolution of product upon addition of water. A detailed account of how to obtain such a cake structure is beyond the scope of the current chapter, but is available in several previous reviews (see above). For the current purposes it is sufficient to focus on the impact of formulation composition on avoiding collapse or meltback. When a product is frozen, the protein and amorphous excipients (e.g., sucrose) are dispersed between ice crystals and any excipient used as a crystalline bulking agent (e.g., glycine). To obtain an appropriate cake structure during lyophilization, the product temperature during primary drying, when the water in ice is sublimed, must be below the characteristic collapse and eutectic melting temperatures of amorphous and crystalline solutes, respectively. Above the eutectic temperature, the melting of crystalline solutes leads to massive loss of porous structure and macroscopic dissolution of the frozen matrix into a "puddle". Above the collapse temperature, which closely coincides with the glass transition temperature (T_g') of the maximally freeze-concentrated amorphous phase, the amorphous phase cannot support its own weight. The result is also a loss of pore structure and a macroscopic shrinkage or collapse of the cake.

It should be noted that the T_g' thermal event, which can be measured with differential scanning calorimetry, is also referred to as a softening event (T_s), rather than the actual glass transition temperature of the freeze-concentrated amorphous phase (Shalaev and Franks, 1995). Whatever the exact nature of the thermal event, it can be detected with differential scanning calorimetry as a second order increase in the baseline of the thermogram, which usually occurs just prior to the onset of the endotherm for the melting of ice (e.g., Her and Nail, 1994). The transition can also be measured with electrothermal methods (Her et al., 1994).

Another powerful method, which is essential for rational development of lyophilized formulations, for determining collapse and eutectic melting temperatures is freeze-drying microscopy (e.g., Nail et al., 1994). With this approach, the formulation of interest is directly examined visually for its performance during a simulated freezing, annealing and drying cycle. All of the critical phase changes, including ice formation, solute crystallization, eutectic melting and collapse, can be detected easily, and the temperature of their occurrence can be measured accurately.

Formulation composition dictates collapse temperature. Each pure amorphous excipient has a characteristic T_g' and collapse temperature; the collapse temperature for the formulation is the mass averaged temperatures of all of the components in the amorphous phase. It is important to design a formulation with maximum collapse temperature, because the rate of drying is directly proportional to the sample temperature during lyophilization. To allow for a reasonable drying time the T_g' should not be lower than –40° C. The T_g' values for pure sucrose is –32° C, while that of pure trehalose is –30° C (Skrabanja et al., 1994; Chang and

Randall, 1992). In contrast, glucose, which should also be avoided because it is a reducing sugar, has a T_g' of –43° C. The collapse temperature of pure protein is about –10° C, which means increasing the protein: sugar mass ratio will increase collapse temperature. Finally, collapse temperature will be decreased if salts and excipients are not maximally crystallized. For example, glycine has a T_g' of –37° C, and its contribution to the amorphous phase can reduce collapse temperature to impractically low values (Carpenter and Chang, 1996). This problem can be avoided by using the appropriate annealing (see Carpenter and Chang, 1996) of the frozen product to maximize crystallization.

Meltback can be avoided by using crystalline solutes with relatively high eutectic melting temperatures. For example, eutectic melting temperatures for mannitol and glycine are –1 and –4° C, respectively. In contrast, additives such as calcium chloride have very low eutectic melting temperatures (e.g., –51° C for calcium chloride). If mannitol or glycine is used as a crystalline bulking agent, the T_g' of the amorphous excipient phase, which is lower that the eutectic melting temperatures of these excipients, will be the value that dictates the temperature of primary drying.

Collapse can also occur during secondary drying, when unfrozen water is desorbed, if the temperature is increased too rapidly. As water is removed from the amorphous phase the T_g of this phase increases. Thus, product temperature can be raised gradually and collapse can be avoided if product temperature at a given time point does not exceed the T_g. Directions for optimizing secondary drying can be found in previous reviews mentioned above.

Steps Taken to Minimize Specific Routes of Protein Chemical Degradation

It is essential that the major routes of chemical, as well as physical, degradation be characterized carefully for each protein, because all of the criteria listed above can be met and a protein might still be damaged during long term storage in the dried solid. For example, methionine oxidation is a common degradation pathway for therapeutic proteins (Manning et al., 1989; Ahern and Manning, 1992; Cleland et al., 1993; Goolcharran et al., 2000). Even in a formulation that prevents protein unfolding and has a T_g exceeding the storage temperature methionine oxidation can proceed at an unacceptably rapid rate. If a methionine residue is on the surface of the protein, maintaining native structure would not prevent this residue's exposure to reactive oxygen species in the dried solid. But why would the glassy matrix not sufficiently retard the mobility of the reactive species to prevent the reaction from occurring? On pharmaceutical time scales (i.e., of many months) relevant motion in a glass is not arrested unless the storage temperature is about 50° C below the T_g (e.g., Hancock et al., 1995; Pikal, 1994; Pikal, 1999). Therefore, the mobility of a relatively small reactive oxygen species and the oxidation of methionine residues are probably not coupled to the glass transition of the formulation. An intriguing alternative, but not mutually exclusive, explanation has been suggested by Steve Prestrelski (personal communication). He proposes that reactive oxygen species accumulating during long-term storage might not be causing damage just in the dried solid. Rather upon rehydration oxygen radicals rapidly react with the protein. This is an important area for future research.

Fortunately, despite the rather complicated and poorly understood theoretical aspects of unacceptable methionine oxidation rates in dried protein formulations, from a practical viewpoint dealing with the problem is relatively straightforward. Methionine oxidation can be minimized by sealing vials under nitrogen and/or using formulation additives (e.g., free methionine) to compete with protein residues for reactive oxygen species. These approaches should be effective whether the oxidation of residues occurs in the dried solid, during rehydration or at both times.

RATIONAL DESIGN OF STABLE LYOPHIILIZED FORMULATIONS

Two critical practical issues need to be considered when choosing excipients for a therapeutic protein formulation. 1) From a financial standpoint, scientists should focus on using excipients and processing approaches that do not need to be licensed from a patent holder. Acceptable protein stability can be achieved readily using excipients and processing methods that are well known to someone, to use the legal term, "skilled in the art of protein formulation." Clearly, however, the specific applications of such compounds and processes to a given protein drug product and/or a class of proteins often can be patented. Such patents may be critical to the company developing that product.

2) From a regulatory standpoint (and financial perspective, also), the formulation scientist should choose from among excipients that are already used in approved parenteral products. For protein stabilizers, the best choices are the disaccharides, sucrose and trehalose. For bulking agents, the best choices are glycine, mannitol and hydroxyethyl starch. For surfactants, usually the Tweens (20, 40 or 80) are preferred.

Based on these practical concerns and the criteria for a successful lyophilized protein product (Table 1), prototypic rational formulations are presented in Table 2. If the formulations suggested are employed, and the criteria in Table 1 are met, then most likely the product will have acceptable stability.

Table 2.
Example of Rational Lyophilized Protein Formulation

1. Buffer that does not acidify during freezing (e.g., Tris, histidine, citrate)
2. Specific ligands/pH that optimizes thermodynamic stability of protein
3. Trehalose or sucrose to inhibit protein unfolding and provide glassy matrix
4. Bulking agent (e.g., mannitol, glycine or hydroxyethyl starch)
5. Nonionic surfactant to reduce protein aggregation

It is important to emphasize that not all components listed in Table 2 may be needed for a given protein product. For example, it may not be necessary to include a nonionic surfactant to aid in reducing protein aggregation. If there is not clear evidence that a given component is beneficial for a formulation, then that component should not be included in the formulation. Also, some components that are used because of "tradition" and/or for purposes such as final formulation tonicity can cause great difficulties in lyophilization. The most common example is the use of NaCl as a tonicity modifier. During lyophilization, NaCl can greatly reduce the collapse temperature of a formulation, if a fraction of the salt does not crystallize (Her et al., 1994). Crystallization of NaCl during freezing and annealing can be inhibited by other excipients (e.g., bulking agents and stabilizing sugars). Because of the low collapse temperature, a low temperature cycle must be used, which increases production time and costs. Also, even if formulations with NaCl can be lyophilized successfully in small research lyophilizers, there is great risk that a large fraction of vials will collapse during large scale manufacturing runs. Thus, if at all possible, NaCl should not be used in lyophilized formulations. Alternative tonicity modifiers include mannitol and glycine, which can also serve as crystalline bulking agents.

Every protein and product has unique characteristics, some of which may cause difficulty in designing stable formulations. Thus, the suggestions in Table 1 should be considered only as a good starting point in formulation development. For the remainder of this chapter we will discuss the rationale for the choice of each of the formulation components, their mechanisms of action and other practical approaches that can be used to increase protein stability.

Choice of Buffer

In terms of protein stability, the main concern with choice of buffer is the potential of certain buffer salts to precipitate during freezing and cause large changes in pH. For example, crystallization of the dibasic form of sodium phosphate results can result in a pH < 4 (van den Burg, 1959; van den Burg and Rose, 1959; Anchordoquy and Carpenter, 1996). Thus, whenever possible, sodium phosphate buffer should be avoided. Although somewhat obvious, it is important to realize that a sodium phosphate system will be present if one starts with potassium phosphate buffer salts and NaCl, as is the case with phosphate buffered saline. Alternative buffers that do not appear to have major pH changes during freezing include Tris and histidine (Anchordoquy and Carpenter, 1996).

A clear example of the detrimental effects of buffer acidification on the stability of a lyophilized protein formulation has been reported by Lam et al. (1996). These researchers documented that a succinate buffer with an initial pH of 5 exhibited a pH drop of 1–2 units during lyophilization. The result was a reduced stability of interferon-γ during storage of the lyophilized formulation, relative to that noted for formulations prepared with glycolate buffer, which does not undergo freezing-induced acidification.

Specific Ligands/pH that Optimizes Thermodynamic Stability of Protein

Often experience gained during purification, intermediate storage protocols and preformulation studies can provide great insight into specific solution conditions (e.g., pH, ionic strength, buffer type) that confer the greatest stability to a given protein. Before choosing the appropriate "general" stabilizers, which are effective at protecting most proteins, it is absolutely essential that the formulation be optimized for the specific factors that increase the physical and chemical stability of a given protein. For example, simply avoiding extremes in pH can drastically reduce the rate of deamidation (Manning et al., 1989; Goolcharran et al., 2000). Also, specific ligands that increase protein physical stability (e.g., by increasing the free energy of unfolding) should be investigated. The stabilizing effects of heparin and other polyanions on growth factors (e.g., Chen et al., 1994; Volkin and Middaugh, 1996) and calcium on DNase (Chen et al., 1999) provide good examples. As will be described below, increasing the free energy of unfolding directly correlates with increased resistance of proteins to denaturation during freezing. In addition, at least in the case of DNase, there was also dramatic increases in storage stability of the lyophilized protein in formulation containing calcium, which were not due to increased structural stabilization during lyophilization (Chen et al., 1999). The mechanism for this effect is unknown.

Moreover, sometimes protein unfolding during freeze-drying can be minimized by optimizing initial solution pH (e.g., Prestrelski et al., 1995). However, in other cases it has been found that initial solution pH does not impact the degree of unfolding arising during lyophilization (Costantino et al., 1995; Carrasquillo et al., 2000). For each protein it is necessary to investigate the effect of initial pH (and potential pH changes during freezing) on structural retention during lyophilization. Infrared spectroscopy can be used to monitor secondary structure in the dried solid, and the resulting data can be used to choose the initial pH that results in the most native-like secondary structure in the dried solid.

Trehalose or Sucrose to Inhibit Protein Unfolding and Provide Glassy Matrix

In this section we will discuss the rationale for choice of stabilizing excipients. Then we will describe the mechanisms for protection of proteins by additives during freezing and drying. Mechanistic insight is important for a clear understanding of protein stabilization during lyophilization and for guiding critical practical choices, such as determining the level of disaccharide needed for optimal protein protection. Finally, we will address this and other practical issues in the use of stabilizing excipients to inhibit protein unfolding during freezing and drying.

Rationale for Choice of Stabilizing Sugar. Among the numerous compounds tested, it appears that the most effective stabilizers of proteins during lyophilization are disaccharides (reviewed in Carpenter et al., 1999). Trehalose and sucrose are the best choices for a stabilizing disaccharide for therapeutic proteins. Both sugars: 1) protect proteins during both freezing and dehydration; 2) are nonreducing; 3) tend to remain amorphous during lyophilization; and 4) have been used in approved parenteral therapeutic products. There are, however, some important differences in the physicochemical properties of these sugars. 1) Trehalose has a higher T_g at a given moisture content than sucrose and, thus, for formulations containing trehalose it may be easier to obtain an appropriate cake structure with an economical lyophilization cycle (Crowe et al., 1998). However, a skilled process engineer should be able to design economical, effective cycles for formulations containing either sugar. In addition, the condition of having a T_g greater than the product storage temperature will hold at higher residual water contents for trehalose. In products with a relatively high protein concentration, the protein could contribute to an increased T_g, which serves to minimize the advantages of trehalose. 2) Trehalose is more resistant than sucrose to acid hydrolysis. Hydrolysis of these disaccharides produces reducing sugars, which must be avoided. Usually this is not a problem, unless very low pH's (ca. < 4) are employed. It should be noted that acid catalyzed hydrolysis of sucrose can occur even in lyophilized solids (Shalaev et al., 2000), suggesting that buffer acidification during freezing (see below) could ultimately result in formation of reducing sugars and resulting damage to proteins in a lyophilized formulation. 3)

Sucrose appears to be somewhat more effective at inhibiting unfolding during lyophilization (Allison et al., 1999). This difference has been most obvious when there is a relatively high protein concentration and a need to employ a relatively high initial concentration of sugar. Evidence to date indicates that less effective stabilization by trehalose is due to the greater propensity of this sugar to phase separate from polymers (Isutzu et al., 1996) and proteins (S.D. Allison, T.W. Randolph, B.S. Chang and J.F. Carpenter, unpublished observation) during freezing and drying. Whether or not this is a problem with a given formulation cannot be predicted. Hence, the relative capacities of sucrose versus trehalose to inhibit lyophilization-induced unfolding of must be examined for each protein. The final choice of sugar for a given protein product should be based on a direct comparison of sucrose and trehalose as stabilizers during the lyophilization cycle and storage in the dried solid.

Reducing sugars, such as maltose or lactose, should not be used. These compounds may effectively inhibit protein unfolding during the lyophilization cycle, but during storage in the dried solid they can degrade proteins via the Maillard reaction between carbonyls of the sugar and free amino groups on the protein (Hageman, 1992; Li et al., 1996).

Mechanism for Freezing Protection. Many different compounds, such as sugars, polyols, certain amino acids, methylamines and salting-out salts, nonspecifically stabilize proteins during freezing or freeze-thawing ("cryoprotection"). The same compounds have also been shown in nonfrozen aqueous systems to increase protein thermodynamic stability (e.g., increase resistance to temperature- or chaotrope-induced unfolding). Protection of proteins by these compounds during freezing and freeze-thawing is due to the same universal thermodynamic mechanism that Timasheff and colleagues have defined for solute-induced protein stabilization in nonfrozen aqueous solution (reviewed in Carpenter and Crowe, 1988; Carpenter and Chang, 1996; Carpenter et al., 1999; Timasheff, 1998).

Usually relatively high concentrations (ca. $> 0.3\,M$) of solute are needed to stabilize proteins because the interactions of the solute with the protein are relatively weak. Stabilizing solutes are excluded preferentially from the surface of the protein. This statement does not mean that no solute molecules bind to the surface of the protein, but rather that there is a lower concentration of solute in the immediate vicinity of the protein than in the bulk solution. The chemical potential of the protein is increased in the presence of a preferentially excluded solute, which is the basis for the effect of solute of protein thermodynamic stability. Considering a two-state model for protein folding ($N \leftarrow \rightarrow D$), the native state is favored thermodynamically (e.g., under physiological conditions) because it has a lower free energy than the denatured state. The key to increasing the thermodynamic stability of the native state is to increase the free energy barrier between the native and denatured states. So how can increasing the chemical potential of the native state result in stabilization? The degree of preferential exclusion and concomitant increase in protein chemical potential correlates directly with the protein surface area. The denatured state has a much greater surface area than the native state. Thus, in the presence of a preferentially excluded solute, the magnitude for the increase in protein chemical potential will be much greater for the denatured than that for the native state; the native state will be stabilized.

This same mechanism applies to the inhibition of smaller scale structural expansions of the native state, which may be sufficient to promote irreversible protein aggregation, by compounds such as sucrose (e.g., Chang et al., 1996a; Kendrick et al., 1997, 1998; Kim et al., 2000). Many proteins are known to form non-native aggregates from species with conformations that are not greatly different from the most compact native state. One example is the molten globule that has perturbed tertiary structure, but native secondary structure. However, even expanded species within that native state ensemble can participate in intermolecular interactions leading to non-native protein aggregates (e.g., Chang et al., 1996a; Kendrick et al., 1997, 1998; Kim et al., 2000). Preferentially excluded solutes shift the equilibrium between protein species towards that with the lowest surface area, i.e., the most compact species in the native state ensemble. As a result aggregation is inhibited.

Timasheff's preferential interaction mechanism also explains the influence of solutes on the degree of assembly of multimeric proteins. Preferentially excluded solutes tend to induce polymerization and stabilize native oligomers since the formation of contact sites between constituent monomers serves to reduce the surface area of the protein exposed to the solvent. Polymerization reduces the thermodynamically unfavorable effect of preferential solute exclusion. In this case, the assembled protein is much more stable than the constituent monomer because of the direct solute effects, but also because of the increase in oligomer stability gained by intersubunit contacts (e.g., Neet and Timm, 1994). In one example of this effect during lyophilization, maintenance of the native tetramer during freezing has been shown to increase the resistance of lactate dehydrogenase to dissociation and inactivation during subsequent drying (Anchordoquy and Carpenter, 1996; Anchordoquy et al., 2001).

How do we know that the preferential exclusion mechanism is actually operative in the frozen state? It is not possible to measure directly the protein-solute interaction in the frozen state. However, the effects of solutes on protein chemical potential and the resulting protein stabilization can be inferred from a freezing study with hemoglobin, in which the protein was partitioned into separate polyethylene glycol-rich and dextran-rich solution phases (Heller et al., 1996). In a phase separated system, thermodynamic equilibrium requires that chemical potential of each component be equal in both phases. Hence, there should be equivalent effects of the different solute conditions in the two phases on the increase in free energy of protein unfolding. In other words, the impact of preferentially excluded solutes on protein stability is equal in the two phases. As predicted from the preferential exclusion mechanism, the degree of structural protection of hemoglobin during freezing (as observed directly in the frozen state with infrared spectroscopy) was equivalent in both phases (Heller et al., 1996).

Mechanism for Inhibition of Dehydration-Induced Unfolding. The interaction of a protein's residues with water are intimately involved with the formation of the native, globular protein structure, and if the protein is dehydrated it will unfold. It has been documented by numerous studies that sucrose and trehalose prevent dehydration-induced unfolding by hydrogen bonding to the dried protein in place of the lost water (e.g., Carpenter and Crowe, 1989; Prestrelski et al., 1993a; Allison et al., 1999; Costantino et al., 1998; Wolkers et al., 1998; Tzannis and Prestrelski, 1999). This so called "water replacement mechanism" is supported by several different observations.

For example, with infrared spectroscopy, it has been found that the band at $1583\,cm^{-1}$ in the spectrum for lysozyme, which is due to hydrogen bonding of water to carboxylate groups, is not present in the spectrum for the dried protein (Carpenter and Crowe, 1989; Remmele et al., 1997; Allison et al., 1999). When lysozyme is dried in the presence of trehalose or sucrose, the carboxylate band is retained in the dried sample, indicating that the sugar is hydrogen bonding in the place of water. Similar results have been obtained with α-lactalbumin and sucrose (Prestrelski et al., 1993a). The magnitude of the retention of the carboxylate band correlates directly with the level of trehalose or sucrose in the lyophilized formulation, as well as with the degree of inhibition of unfolding (Allison et al., 1999). These effects of sugars on proteins in the dried solid are not due to the presence of increased amounts of water in the formulations dried with sugars (Allison et al., 1999; Tzannis and Prestrelski, 1999). The level of water in formulations dried with the sugars is as low as that for the protein lyophilized from just buffer or water, and is so low that the hydration shell of the protein is essentially completely removed (Prestrelski et al., 1993a; Krielgaard et al., 1998a, 1999; Allison et al., 1999).

Furthermore, Farhat et al., (1998) used infrared spectroscopy to study dried gelatin-sugar mixtures and the hydration behaviors of the mixtures, and concluded that the sugars hydrogen bond to the dried protein. Costantino et al. (1997) studied water sorption of proteins lyophilized with sucrose or trehalose and found that the solid-state interactions between protein and sugar reduced the availability of water binding sites. Tzannis and Prestrelski (1999) found that water sorption behavior of spray-dried protein formulations containing stabilizing levels of sucrose indicated hydrogen bonding between dried protein and the sugar. However, if an excessively high sucrose:protein ratio was employed, due to formation of protein- and sugar-rich phases, hydrogen bonding of sucrose to protein and resulting protein stabilization were reduced. Also, Suzuki et al. (1998; 1999) found that there was a high degree of stabilization of lactate dehydrogenase when sucrose remained amorphous and hydrogen bonded to the dried protein. Crystallization of sucrose abolished hydrogen bonding between the sugar and dried protein, as well as protein stabilization.

Further support for the water replacement mechanisms comes from two studies that compared the relative effectiveness of saccharides of increasing molecular weight for inhibiting lyophilization-induced unfolding. Tanaka and colleagues (1991) found that the capacity to protect catalase during freeze-drying was inversely related to saccharide size. They suggested that as the size of the saccharide increases, steric hindrance interferes with hydrogen bonding between the saccharide and the dried protein. In support of this contention, the carboxylate band is only minimally detectable in the infrared spectrum of lysozyme freeze-dried in the presence of glassy dextran and the protein is unfolded (Allison et al., 1999). With infrared spectroscopy, Prestrelski et al. (1995) found that as the molecular weight of a carbohydrate additive was increased the capacity to inhibit unfolding of interleukin-2 during lyophilization decreased, and the level of protein aggregation after rehydration increased. Also, it was clear that protection of the protein did not correlate directly with the formation of a glass (all samples were found to be amorphous) or with the glass transition temperature of the sample (the T_g increased as carbohydrate molecular weight increased). Rather, there was a negative correlation between stabilization and molecular weight, which is to be expected if protection during drying is due to the water replacement mechanism.

An alternative mechanism for stabilization of proteins during dehydration states that proteins are simply mechanically immobilized in a glassy, solid matrix during dehydration (e.g., Franks, 1991). The restriction of translational and relaxation processes is thought to inhibit protein unfolding, and spatial separation between protein molecules (i.e., "dilution" of protein molecules within the glassy matrix) is proposed to prevent aggregation. It is clear that protective additives must partition with the protein into the amorphous phase of the dried sample. If the compound crystallizes during lyophilization it does not inhibit protein unfolding (Carpenter et al., 1993; Izutsu et al., 1993; Krielgaard et al., 1999). Also, spatially separating protein molecules can help favor refolding over aggregation during rehydration (Allison et al., 1998, 2000). However, as evidenced by the failure of glassy dextran matrices to prevent unfolding (Prestrelski et al., 1995; Krielgaard et al., 1999; Allison et al., 1998, 1999, 2000), simply forming a glassy solid is not sufficient for inhibiting dehydration-induced unfolding.

Practical Considerations. The protein mass in a unit dose (or multidose) vial of therapeutic protein is dictated by the amount to be given to the patient.

With a lyophilized product, which can be reconstituted to a volume different from the prelyophilization volume, the initial protein concentration can be varied without changing the total protein mass. Changes in initial protein concentration can affect relative loss of protein due to interaction with vial surfaces (e.g., Page et al., 2000), protein stability during processing, the collapse temperature of the formulation, the T_g of the final dried formulation and the lyophilization process itself. Thus, protein concentration can be an important variable to investigate during formulation development.

Increasing initial protein concentration leads to apparent increased resistance to denaturation during freezing (Strambini and Gabellieri, 1996; Chang et al., 1996c; Krielgaard et al., 1998b). This phenomenon can be demonstrated by determining the percentage protein aggregated after freeze-thawing, which varies inversely with protein concentration. Increasing protein concentration can directly reduce freezing-induced protein unfolding because one component of protein damage during freezing appears to involve protein denaturation during formation of the ice-water interface (Strambini and Gabellieri, 1996; Chang et al., 1996c; Krielgaard et al., 1998b). Assuming that only a finite number of protein molecules can be unfolded per unit area at this interface (Krielgaard et al., 1998b), increasing the initial protein concentration will lead to a smaller percentage of damaged molecules. Of course, other factors such as direct cold denaturation and freeze concentration of salts may predominate over surface denaturation during freezing. For practical purposes, it is not necessary to sort out the relative contribution of the various stresses to freezing-induced damage. Rather, is important to include realative high protein concentrations during early formulation development in order to increase the "intrinsic" resistance of the protein to denaturation. The greater the intrinsic stability of the protein, the lower the concentration of excipient that will be needed.

Another advantage of employing a relatively high initial protein concentration is that increasing protein : disaccharide mass ratio increases both formulation collapse temperature and T_g of the final dried product. In addition, the formulation volume will be reduced, which will reduce the duration and costs of lyophilization.

Of course, if the ratio of excipient to protein is not sufficiently high, there will not be adequate stabilization of the protein during freezing and dehydration. Freezing protection depends on the initial bulk concentration of the sugar, and if the given protein is freeze-labile sometimes concentrations exceeding 5% (wt/vol) are needed to maximize stabilization. However, often the stability of the protein is sufficient during freezing that the level of sugar needed for optimal protection during lyophilization is dictated by that required for inhibition of unfolding during dehydration. Protection during drying depends on the final mass ratio between the sugar and the protein (Pikal, 1994; Carpenter and Chang, 1996;

Cleland et al., 2001). Recently it has been demonstrated that the sugar: protein mass ratio predicts not only degree of native structural retention during lyophilization, but also the long-term storage stability of a lyophilized protein (Cleland et al., 2001). Generally, a weight ratio of sugar to protein of at least 1 : 1 is required for inhibiting lyophilization-induced unfolding, with optimal stability being reached at around 3–5 : 1. In practice, with the protein concentration held constant, a range of sugar concentrations can be tested during formulation screening to discern the optimal concentration needed for retention of native protein in the dried solid. In general, the optimal sugar concentration for stabilizing the protein during lyophilization will also provide storage stability, if the final dried powder has a T_g well above the storage temperature. Finally, it should be stressed that the minimal effective level of sugar should be employed, because of the effect of changing protein : excipient ratio on glass transition temperatures. Also, if excess levels of sugar are employed there is an increased risk of sugar crystallization if a product should happen to be held at temperatures near or above the formulation T_g (e.g., Krielgaard et al., 1999).

Bulking Agent (e.g., Mannitol, Glycine or Hydroxyethyl Starch)

If the product has a relatively low mass of protein per vial, often it will necessary to have a bulking agent in the formulation to prevent the protein from being lost from the vial during drying and to form the product cake. Mannitol and glycine are examples of bulking agents, which can also serve as tonicity modifiers, that usually crystallize to a substantial degree during lyophilization (Pikal, 1994; Carpenter and Chang, 1996). A disaccharide protein stabilizer will need to be used in combination with these bulking agents. One drawback of mannitol and glycine is that often an annealing step is required to assure maximum crystallization. If a fraction of the bulking agent remains amorphous during lyophilization, there may be problems with obtaining a suitable cake structure, as well as a risk of excipient crystallization during subsequent storage in the dried solid (e.g., Carpenter and Chang, 1996; Carpenter et al., 1997). An alternative, amophous bulking agent is the polymer hydroxyethyl starch, which is used clinically as a plasma volume expander. Its main advantages in lyophilization are that it has a high collapse temperature (ca. −10°C), it forms strong cakes and it has a relatively high T_g for a given moisture content (e.g., >200°C at 2% residual moisture). However, as is the case with dextran, the large polymer hydroxyethyl starch does not inhibit protein unfolding during dehydration, and may actually foster additional protein unfolding due to phase separation from the protein during freezing and drying. Again, stabilizing disaccharides must be used with hydroxyethyl starch to inhibit lyophilization-induced protein unfolding.

Nonionic Surfactant to Inhibit Aggregation

Surfactants are often included in protein formulations to minimize interfacial denaturation and aggregation, e.g., at vial surfaces, due to bubble entrainment during filling or because of agitation (Chang et al., 1996c; Krielgaard et al., 1998b; Page et al., 2000). For a lyophilized product, a surfactant can be beneficial to minimize protein aggregation during vial filling, freezing and rehydration. Generally, a surfactant will not inhibit protein unfolding during dehydration (Krielgaard et al., 1998a). The mechanisms by which surfactants inhibit protein aggregation will be described in a separate chapter in this volume. For the current discussion it is sufficient to stress that a surfactant should not be included in a lyophilized product, unless there is direct evidence that increases recovery of native protein in the rehydrated sample. Surfactants can decrease the free energy of unfolding of some proteins, which may cause the compound to actually foster aggregation. Also, they have been shown to inhibit the assembly of small soluble aggregates into higher order soluble oligomers and insoluble aggregates (e.g., Krielgaard et al., 1998b). This effect can manifest itself during lyophilization and rehydration. As a result, the presence of a surfactant can cause an undesirable increase in the level of soluble aggregates.

ACKNOWLEDGMENTS

We gratefully acknowledge support from grants from the National Science Foundation (BES 9816975), Amgen, Inc., Genentech, Inc., Genetics Institute, Inc. and Genencor International. We also thank the National Science Foundation, the US Department of Education, the American Foundation for Pharmaceutical Education, the Pharmaceutical Manufacturer and Research Association and the Colorado Institute for Research in Biotechnology for providing predoctoral fellowships to our graduate students. WGR is the recipient of a postdoctoral fellowship from NIH (CA79446–02).

REFERENCES

Ahern, T.J., and Manning, M.C., 1992. *Stability of Protein Pharmaceuticals: Part. A. Chemical and Physical Pathways of Protein Degradation*, Plenum Press, New York.

Allison, S.D., Dong, A., and Carpenter, J.F., 1996. Counteracting effects of thiocyanate and sucrose on chymotrypsinogen secondary structure and aggregation during freezing, drying and rehydration. *Biophys. J.* 71:2022.

Allison, S.D., Randolph, T.W., Davis, A., Middleton, K., and Carpenter, J.F., 1998. Effects of drying methods and additives on structure and function of actin: Mechanisms of dehydration-induced damage and its inhibition. *Arch. Biochem. Biophys.* 358: 171.

Allison, S.D., Randolph, T.W., Chang, B.S., and Carpenter, J.F., 1999. Hydrogen bonding between sugar and protein is responsible for inhibiting dehydration-induced protein unfolding. *Arch. Biochem. Biophys.* 365:289.

Allison, S.D., Manning, M.C., Randolph, T.W., Middleton, K., Davis, A., and Carpenter, J.F., 2000. Optimization of storage stability of lyophilized actin using combinations of disaccharides and dextran. *J. Pharm. Sci.* 89:199.

Anchordoguy, T.J., and Carpenter, J.F., 1996. Polymers protect lactate dehydrogenase during freeze-drying by inhibiting dissociation in the frozen state. *Arch. Biochem. Biophys.* 332:231.

Anchordoquy, T.J., Izutsu, K.I., Randolph, T.W., and Carpenter, 2001. Maintenance of quaternary structure in the frozen state stabilizes lactate dehydrogenase during freeze-drying. *Arch. Biochem. Biophys.* (In Press).

Angell, C.A., 1995. Formation of glasses from liquids and biopolymers. *Science* 267:1924.

Arakawa, T., Prestrelski, S., Kinney, W., and Carpenter, J.F., 1993. Factors affecting short-term and long-term stabilities of proteins. *Adv. Drug Delivery Rev.* 10:1.

Bell, J.A., 1999. X-ray crystal structures of a severely desiccated protein. *Prot. Sci.* 8:2033.

Bummer, P.M., and Koppenol, S., 2000. Chemical and physical considerations in protein and peptide stability, in: *Protein Formulation and Delivery*, E.J. McNally, ed., Marcel Dekker, New York.

Brange, J., 2000. Physical stability of proteins, in: *Pharmaceutical Formulation Development of Peptides and Proteins*, S. Frokjaer and L. Hovgaard, eds., Taylor and Francis, London.

Cappola, M.L., 2000. Freeze-drying concepts: The basics, in: *Protein Formulation and Delivery*, E.J. McNally, ed., Marcel Dekker, New York.

Carpenter, J.F., and Chang, B.S., 1996. Lyophilization of protein pharmaceuticals, in: *Biotechnology and Biopharmaceutical Manufacturing, Processing, and Preservation*, K.E. Avis and V.L. Wu, eds., Interpharm Press, Buffalo Grove, IL.

Carpenter, J.F., Izutsu, K., and Randolph, T.W., 1999. Freezing- and drying-induced structural changes in proteins and their inhibition by stabilizing additives. *Pharmaceutical Freeze-Drying*, L. Rey and J.C. May, eds., Marcel Dekker, New York.

Carpenter, J.F., Pikal, M.J., Chang, B.S., and Randolph, T.W., 1997. Rational design of stable lyophilized protein formulations: Some practical advice. *Pharm. Res.* 14:969–975

Carpenter, J.F., Prestrelski, S., and Arakawa, T., 1993. Separation of freezing- and drying-induced denaturation of lyophilized proteins by stress-specific stabilization: I. Enzyme activity and calorimetric studies. *Arch. Biochem. Biophys.* 303:456.

Carpenter, J.F., Prestrelski, S.J., and Dong, A., 1998. Application of infrared spectroscopy to development of stable lyophilized protein formulations. *Eur. J. Pharm. Biopharm.* 45:231.

Carpenter, J.F., and Crowe, J.H., 1988. The mechanism of cryoprotection of proteins by solutes. *Cryobiology* 25:244.

Carpenter, J.F., and Crowe, 1989. Infrared spectroscopic studies on the interaction of carbohydrates with dried proteins. *Biochemistry* 28:3916.

Chang, B.S., and Randall, C.S., 1992. Use of subambient thermal analysis to optimize protein lyophilization. *Cryobiology* 29:632.

Chang, B.S., and Fisher, N.L., 1995. Development of an efficient single-step freeze-drying cycle for protein formulations. *Pharm. Res.* 12:831.

Chang, B.S., Beauvais, R.M., Arakawa, T., Narhi, L.O., Dong, A., Aparisio, D.I., and Carpenter, J.F., 1996a. Formation of an active dimer during storage of interleukin-1 receptor antagonist in aqueous solution. *Biophys. J.* 71:3399.

Chang, B.S., Beauvais, R.M., and Carpenter, J.F., 1996b. Physical factors affecting the storage stability of freeze-dried interleukin-1 receptor antagonist: Glass transition and protein conformation. *Arch. Biochem. Biophys.* 331:249.

Chang, B.S., Kendrick, B.S., and Carpenter, J.F., 1996c. Surface-induced denaturation of proteins during freezing and its inhibition by surfactants. *J. Pharm. Sci.* 85:1325.

Chang, B.S., Kendrick, B.S., Carpenter, J.F. 1996c. Surface-induced denaturation of proteins during freezing and its inhibition by surfactants. *J. Pharm. Sci.* 85:1325.

Chen, B-L., Arakawa, T., Hsu, L., Narhi, L., Tressel, T.J., and Chen, S.L., 1994. Strategies to suppress aggregation of recombinant keratinocyte growth factor during liquid formulation development. *J. Pharm. Sci.* 83:1657.

Chen, B., Costantino, H.R., Liu, J., Hsu, C.C., and Shire, S.J., 1999. Influence of calcium ions on the structure and stability of recombinant human deoxyribonuclease I in the aqueous and lyophilized states. *J. Pharm. Sci.* 88:477.

Cleland, J.L., Powell, M.F., and Shire, S.J., 1993. The development of stable protein formulations—A close look at protein aggregation, deamidation and oxidation. *Crit. Rev. Ther. Drug* 11:60.

Cleland, J.L., Lam, X., Kendrick, B.S., Yang, J., Yang, T-Z., Overcashier, D., Brooks, D., Hsu, C., and Carpenter, J.F., 2001. A specific molar ratio of stabilizer to protein is required for storage stability of a lyophilized monoclonal antibody. *J. Pharm. Sci.* 90:310.

Carrasquillo, K.G., Sanchez, C., Griebenow, K., 2000. Relationship between conformational stability and lyophilization-induced structural changes in chymotrypsin. *Biotechnol. Appl. Biochem.* 31:41.

Costantino, H.R., Curley, J.G., Wu, S., and Hsu, C.C., 1997. Water sorption behavior of lyophilized protein-sugar systems and implications for solid-state interactions. *Int. J. Pharm.* 166:211.

Costantino, H.R., Griebenow, K., Mishra, P., Langer, R., and Klibanov, A.M., 1995. Fourier-transform infrared spectroscopic investigation of protein stability in the lyophilized form. *Biochim. Biophys. Acta* 1253:69.

Costantino, H.R., Carrasquillo, K.G., Cordero, R.A., Mumenthaler, M., Hsu, C.C., and Griebenow, K., 1998. Effect of excipients on the stability and structure of lyophilized recombinant human growth hormone. *J. Pharm. Sci.* 87:1412.

Craig, D.Q., Royall, P.G., 1998. The use of modulated temperature DSC for the study of pharmaceutical systems: potential uses and limitations. *Pharm. Res.* 15:1152.

Crowe, J.H., Carpenter, J.F., and Crowe, L.M., 1998. The role of vitrification in anhydrobiosis. *Annu. Rev. Physiol.* 60:73.

DeGrazio, F., Flynn, K., 1992. Lyophilization closures for protein based drugs. *J. Parenter. Sci. Technol.* 46:54.

Dong, A., Prestrelski, S.J., Allison, S.D., and Carpenter, J.F., 1995. Infrared spectroscopic studies of lyophilization- and temperature-induced protein aggregation. *J. Pharm. Sci.* 84:415.

Duddu, S.P., and Dal Monte, P.R., 1997. Effect of glass transition temperature on the stability of lyophilized formulations containing a chimeric therapeutic monoclonal antibody. *Pharm. Res.* 14:591.

Duddu, S.P., Zhang, G., Dal Monte, P.R., 1997. The relationship between protein aggregation and molecular mobility below the glass transition temperature of lyophilized formulations containing a monoclonal antibody. *Pharm. Res.* 14:596.

Farhat, I.A., Orset, S., Moreau, P., Blanchard, J.M.V., 1998. FTIR study of hydration phenomena in protein-sugar systems. *J. Coll. Interface Sci.* 207:200.

Franks, F., 1990. Freeze-drying: From empiricism to predictability. *Cryo-Letters* 11:93.

Franks, F., Hatley, R.H.M., and Mathias, S.F., 1991. Material science and the production of shelf stable biologicals. *BioPharm* 4(9):38.

Gatlin, L.A., and Nail, S.L., 1994. Protein purification process engineering. Freeze-drying: A practical overview. *Bioprocess Technol.* 18:317.

Goolcharran, C., Khossravi, M., and Borchardt, R.T., 2000. Chemical pathways of protein and peptide degradation, in: *Pharmaceutical Formulation Development of Peptides and Proteins*, S. Frokjaer and L. Hovgaard, eds., Taylor and Francis, London.

Griebenow, K., and Klibanov, A.M., 1995. Lyophilization-induced reversible changes in the secondary structure of proteins. *Proc. Natl. Acad. Sci. USA* 92:10969.

Gu, L.D., Erds, E.A., Chiang, H.-S., Calderwood, T., Tsai, K., Visor, G.C., Duffy, J., Hsu, W.C., and Foster, L.C., 1991. Stability of interleukin-1 beta (IL-1 beta) in aqueous solution: analytical methods, kinetics, products, and solution formulation implications. *Pharm. Res.* 8:485.

Hageman, M., 1992. Water sorption and solid state stability of proteins. in: *Stability of Protein Pharmaceuticals. Part A. Chemical and Physical Pathways of Protein Degradation*, T. Ahern and M.C. Manning, eds., Plenum Press, New York.

Hancock, B.C., Shamblin, S.L., and Zografi, G., 1995. Molecular mobility of amorphous pharmaceutical solids below their glass transition temperature. *Pharm. Res.* 12:799.

Heller, M.C., Carpenter, J.F., and Randolph, T.W., 1996. Effects of phase separating systems on lyophilized hemoglobin. *J. Pharm. Sci.* 85:1358.

Her, L.M., Jefferis, R.P., Gatlin, L.A., Braxton, B., and Nail, S.L., 1994. Measurement of glass transition temperatures in freeze concentrated solutions of non-electrolytes by electrical thermal analysis. *Pharm. Res.* 11:1023.

Her, L.M., and Nail, S.L., 1994. Measurement of glass transition temperatures of freeze-concentrated solutes by differential scanning calorimetry. *Pharm. Res.* 11:54.

Her, L.M., Deras, M., and Nail, S.L., 1995. Electrolyte-induced changes in glass transition temperatures of freeze-concentrated solutes. *Pharm. Res.* 12:768.

Hora, M.S., and Wolfe, S.N., 1999. Critical steps in the preparation of elastomeric closures for biopharmaceutical freeze-dried products, in: *Freeze-Drying/Lyophilization of Pharmaceuticals and Biological Products*, L. Rey and J.C. May, eds., Marcell Dekker, New York.

Izutsu, K., Yoshioka, S., Kojima, S., Randolph, T.W., and Carpenter, J.F., 1996. Effects of sugars and polymers on crystallization of poly (ethylene glycol) in frozen solutions: Phase separation between incompatible polymers. *Pharm. Res.* 13:1393.

Izutsu, K., Yoshioka, S., and Teroa, T., 1993. Decreased protein-stabilizing effects of cryoprotectants due to crystallization. *Pharm. Res.* 10:1232.

Kendrick, B.S., Chang, B.S., Arakawa, T., Peterson, B., Randolph, T.W., Manning, M.C., and Carpenter, J.F., 1997. Preferential exclusion of sucrose from recombinant interleukin-1 receptor antagonist: Role in restricted conformational mobility and compaction of native state. *Proc. Nat. Acad. Sci.* 94:11917.

Kendrick, B.S., Carpenter, J.F., Cleland, J.L., and Randolph, T.W., 1998. A transient expansion of the native state precedes aggregation of recombinant human interferon-gamma. *Proc. Nat. Acad. Sci.* 95:14142.

Kim, Y., Wall, J.S., Meyer, J., Murphy, C., Randolph, T.W., Manning, M.C., Solomon, A., and Carpenter, J.F., 2000. Thermodynamic modulation of light chain amyloid fibril formation. *J. Biol. Chem.* 275:1570.

Kreilgaard, L., Frokjaer, S., Flink, J.M., Randolph, T.W., and Carpenter, J.F., 1998a. Effects of additives on the stability of recombinant human Factor XIII during freeze-drying and storage in the dried solid. *Arch. Biochem. Biophys.* 360:121.

Kreilgaard, L., Jones, L., Randolph, T.W., Frokjaer, S., Flink, J., Manning, M.C., and Carpenter, J.F., 1998b. Effects of Tween 20 on agitation- and freeze-thawing-induced aggregation of recombinant Factor XIII. *J. Pharm. Sci.* 87:1597.

Kreilgaard, L., Frokjaer, S., Flink, J.M., Randolph, T.W., and Carpenter, J.F., 1999. Effects of additives on the stability of *Humicola lanuginosa* lipase during freeze-drying and storage in the dried solid. *J. Pharm. Sci.* 88:281.

Lam, X.M., Costantino, H.R., Overcashier, D.E., Nguyen, T.H., and Hsu, C.C., 1996. Replacing succinate with glycolate buffer improves the stability of lyophilized interferon-gamma. *Int. J. Pharm.* 142:85.

Levine, H., and Slade, L., 1988. Thermomechanical properties of small-carbohydrate-water glasses and "rubbers": Kinetically metastable systems at subzero temperatures. *J. Chem. Soc. Faraday Trans.* 1(84):2619.

Levine, H., and Slade, L., 1992. Glass transitions in foods, in: *Physical Chemistry of Foods*, H.S. Shartxberg and R.W. Hartel, eds., Marcel Dekker, NewYork.

Li, S., Patapoff, T.W., Overcashier, D., Hsu, C., Nguyen, T.H., and Borchardt, R.T., 1996. Effects of reducing sugars on the chemical stability of human relaxin in the lyophilized state. *J. Pharm. Sci.* 85:873.

Lueckel, B., Helk, B., Bodmer, D., and Leuenberger, H., 1998. Effects of formulation and process variables on the aggregation of freeze-dried interleukin-6 (IL-6) after lyophilization and on storage. *Pharm. Dev. Technol.* 3:337.

Manning, M.C., Patel, K., and Borchardt, R.T., 1989. Stability of protein pharmaceuticals. *Pharm. Res.* 6:903.

Nail, S.L., and Gatlin, L.A., 1993. Freeze-drying: Principles and practice, in: *Pharmaceutical Dosage Forms: Parenteral Medications*, K.E. Avis, H.A. Lieberman and L. Lachman, eds., Marcel Dekker, New York.

Nail, S.L., Her, L.M., Proffitt, C.P., and Nail L.L., 1994. An improved microscope stage for direct observation of freezing and freeze drying. *Pharm. Res.* 11:1098.

Neet, K.E., and Timm, D.E., 1994. Conformational stability of dimeric proteins: quantitative studies by equilibrium denaturation. *Prot. Sci.* 3:2167.

Page C., Dawson, P., Woollacott, D., Thorpe, R., Mire-Sluis, 2000. Development of a lyophilization formulation that preserves the biological activity of the platelet-inducing cytokine interleukin-11 at low concentrations. *J. Pharm. Pharmacol.* 52:19.

Pikal, M.J., 1990. Freeze-drying of proteins. Part I: Process design. *BioPharm* 3(8):18.

Pikal, M.J., 1994. Freeze-drying of proteins, in: *Formulation and Delivery of Proteins and Peptides*, J.L. Cleland and R. Langer, eds., Symposium Series 567, American Chemical Society, Washington, DC.

Pikal, M.J., 1999. Mechanisms of protein stabilization during freeze-drying and storage: The relative importance of thermodynamic stabilization and glassy state relaxation dynamics, in: *Freeze-Drying/Lyophilization of Pharmaceuticals and Biological Products*, L. Rey and J.C. May, eds., Marcell Dekker, New York.

Pikal, M.J., and Shah, S., 1992. Moisture transfer from stopper to product and resulting stability implications. *Dev. Biol. Standard.* 74:165.

Pikal, M.J., Shah, S., Roy, M.L., and Putman, R., 1990. The secondary drying stage of freeze-drying: Drying kinetics as a function of temperature and chamber pressure. *Int. J. Pharm.* 60:203.

Prestrelski, S.J., Tedeschi, N., Arakawa, T., and Carpenter, J.F., 1993a. Dehydration-induced conformational changes in proteins and their inhibition by stabilizers. *Biophys. J.* 65:661.

Prestrelski, S.J., Arakawa, T., and Carpenter, J.F., 1993b. Separation of freezing- and drying-induced denaturation of lyophilized proteins by stress-specific stabilization: II. Structural studies using infrared spectroscopy. *Arch. Biochem. Biophys.* 303:465.

Prestrelski, S.J., Pikal, K.A., and Arakawa, T., 1995. Optimization of lyophilization conditions for recombinant interleukin-2 by dried state conformational analysis using Fourier transform infrared spectroscopy. *Pharm. Res.* 12:1250.

Remmele, R.L. Jr., Stushnoff, C., and Carpenter, J.F., 1997. Real-time *in situ* monitoring of lysozyme during lyophilization using infrared spectroscopy: dehydration stress in the presence of sucrose. *Pharm. Res.* 14:1548.

Rey, L., and May, J.C., 1999. *Freeze-drying/Lyophilization of Pharmaceuticals and Biological Products*, Marcel Dekker, New York.

Roy, M.L., Pikal, M.J., Rickard, E.C., and Maloney, A.M., 1991. The effects of formulation and moisture on the stability of a freeze-dried monoclonal antibody-vinca conjugate: A test of the WLF glass transition theory. *Dev. Biol. Standard.* 74:323.

Shalaev, E.Y., and Franks, F., 1995. Structural glass transitions and thermophysical processes in amorphous carbohydrates and their supersaturated solutions. *J. Chem. Soc. Faraday Trans.* 91:1511.

Shalaev, E.Y., Lu, Q., Shalaeva, M., and Zografi, G., 2000. Acid-catalyzed inversion of sucrose in the amorphous state at very low levels of residual water. *Pharm. Res.* 17:366.

Skrabanja, A.T.P., de Meere, A.L.J., de Ruiter Rien, and van der Oetelaar, P.J.M., 1994. Lyophilization of iotechnology products. *PDA J. Pharm. Sci. Technol.* 48:311.

Strambini, G.B., and Gabellieri, E., 1996. Proteins in frozen solutions: Evidence of ice-induced partial unfolding. *Biophys. J.* 70:971.

Suzuki, T., Imamura, K., Fujimoto, H., and Okazaki, M., 1998. Relationship between hermal stabilizing effect of sucrose on LDH and sucrose-LDH hydrogen bond. *J. Chem. Eng. Japan* 31:565.

Suzuki, T., Imamura, K., Fujimoto, H., and Okazaki, M., 1999. Role of sucrose-LDH hydrogen bond for thermal stabilizing effect of sucrose on freeze-dried LDH. *Drying Technol.* 17:1429.

Tanaka, T., Takeda, T., and Miyajama, R., 1991. Cryoprotective effect of saccharides on denaturation of catalase during freeze-drying. *Chem. Pharm. Bull.* 39:1091.

Timasheff, S.N., 1998. Control of protein stability and reactions by weakly interacting cosolvents: The simplicity of the complicated. *Adv. Prot. Chem.* 51:355.

Tzannis, S.T., and Prestrelski, S.J., 1999. Moisture effects on protein-excipient interactions in spray-dried powders. Nature of destabilizing effects of sucrose. *J. Pharm. Sci.* 88:360.

van den Berg, L., 1959. The effect of addition of sodium and potassium chloride to the reciprocal system: KH_2-PO_4-Na_2HPO_4-H_2O on pH and composition during freezing, *Arch. Biochem. Biophys.* 84:305–315.

van den Berg, L., and Rose, D., 1959. The effect of freezing on the pH and composition of sodium and potassium solutions: The reciprocal system KH_2-PO_4-Na_2HPO_4-H_2O. *Arch. Biochem. Biophys.* 81:319.

Verdonck, E., Schaap, K., and Thomas, L.C., 1999. A discussion of the principles and applications of modulated temperature DSC (MTDSC). *Int. J. Pharm.* 192:3.

Volkin, D.B., and Middaugh, C.R., 1996. The characterization, stabilization, and formulation of acidic fibroblast growth factor. *Pharm. Biotechnol.* 9:181.

Wolkers, W.F., van Kilsdonk, M.G., and Hoekstra, F.A., 1998. Dehydration-induced conformational changes of poly-L-lysine as influenced by drying rate and carbohydrates. *Biochim. Biophys. Acta* 1425:127.

Yoshioka, S., Aso, Y., Kojima, S., and Tanimoto, T., 2000. Effect of polymer excipients on the enzyme activity of lyophilized bilirubin oxidase and beta-galactosidase formulations. *Chem. Pharm. Bull.* 48:283.

Yoshioka, S., Aso, Y., and Kojima, S., 1999. The effect of excipients on the molecular mobility of lyophilized formulations, as measured by glass transition temperature and NMR relaxation-based critical mobility temperature. *Pharm. Res.* 16:135.

Zhang, M.Z., Wen, J., Arakawa, T., and Prestrelski S.J., 1995. A new strategy for enhancing the stability of lyophilized protein: The effect of reconstitution medium on keratinocyte growth factor. *Pharm. Res.* 12:1447.

Zhang, M.Z., Pikal, K., Nguyen, T., Arakawa, T., and Prestrelski, S.J., 1996. The effect of the reconstitution medium on aggregation of lyophilized recombinant interleukin-2 and ribonuclease A. *Pharm. Res.* 13:643.

6

SPRAY-DRYING OF PROTEINS

Geoffrey Lee

INTRODUCTION: WHY SPRAY-DRY A PROTEIN?

The answer to this question is simple: spray-drying is an effective, efficient means of producing peptide/protein-loaded powders suitable for pulmonary delivery. In addition, fine powders of proteins have application in other delivery systems (e.g., implantable pumps). If the correct formulation and spray-drying conditions can be identified, then a product can be obtained with a high yield and having a large fine-particle dose. Spray-drying is therefore a realistic alternative to the widespread practice of air-jet milling used to produce powders of low molecular-weight actives for inhalation. Indeed, air-jet milling should probably be avoided with peptides or proteins, because of possible problems of physical instability and inactivation. Apart from these considerations, spray-drying is also a suitable technique for embedding sensitive peptides or proteins in stabilizing excipients ("carriers") such as disaccharides or amino acids.

The spray-drying of an aqueous solution of protein without excipients can lead to unfolding, aggregation and inactivation. For this reason the protein must certainly be formulated with one or more of a number of substances to improve both its process and storage stabilities. The use of glassy carriers such as sucrose, trehalose or arginine is known to reduce the aggregation/inactivation of proteins during spray-drying (e.g., Broadhead et al., 1993; Mumenthaler et al., 1994; Andya et al., 1999). Additionally, these excipients can provide a high degree of

Geoffrey Lee • Department of Pharmaceutical Technology, Friedrich-Alexander University, Erlangen, Germany.

Rational Design of Stable Protein Formulations, edited by Carpenter and Manning. Kluwer Academic / Plenum Publishers, New York, 2002.

protection on storage at temperatures below the glass transition temperature, T_g, of the spray-dried formulation. The addition of surfactants or zinc salts has also been shown to reduce the extent of aggregation of at least one protein during spray-drying (e.g., Maa et al., 1997a, 1998a). The correct selection of effective excipients for a particular protein remains, however, a fortuitous affair. The formulator must consider the complex interplay of spray-drying conditions, T_g and residual water content, possible detrimental effects of surfactant, the yield obtained, particle size and morphology, and the process and storage stabilities of the product.

Anyone considering the use of spray-drying should first consult Master's *Spray-Drying Handbook* (Masters, 1991). This volume contains a great deal of extremely useful information about the theory and practice of spray-drying, and is simply invaluable. It includes, however, no references to the spray-drying of therapeutic (i.e., pharmaceutical) peptides and proteins, since the essential work in this area has been undertaken in the last 10 years. In this chapter, I aim to present a sound, practical approach to the subject, organized as follows. First, there is a brief review of the major developments of the last 10 years. Secondly, a discussion of the practice of spray-drying is presented, centering on equipment and the process conditions used to produce spray-dried, protein-containing particles. Thirdly, the vital influence of the formulation on protein stability and product quality will be discussed at length. Therefore, both the industrial pharmacist and the academic researcher should find the information they need about the spray-drying of proteins for pharmaceutical application.

DEVELOPMENTS IN THE LAST 10 YEARS

Although the spray-drying of sensitive biologicals, such as enzymes and sera, has been routinely practiced for a number of years, the first paper of interest for the current chapter is that of Labrude et al. in 1989 (Labrude et al., 1989). These authors spray-dried oxyhemoglobin and found that its chemical stability during the process could be improved by adding sucrose to the spray solution. A parallel could therefore be drawn with the protein-stabilizing action of carbohydrates during freeze-drying (Carpenter et al., 1994; Pikal, 1994; Franks, 1998). However, further development of this theme remained slow in the early 90's. A review article on the spray-drying of pharmaceuticals by Broadhead et al. (1992) contains therefore only this single reference. Their article is useful, however, since it summarizes some important studies of enzyme inactivation during spray-drying, which is, of course, of direct relevance to pharmaceutical proteins.

Two important pharmaceutical papers appeared in 1994. Broadhead et al. (1993) published a systematic study of the effect of process and formulation variables on the properties of spray-dried β-galactosidase. They demonstrated the effects of spray-drying temperature on product yield and enzyme inactivation. Additionally, they confirmed that stabilizers, such as disaccharides and amino acids, preserve enzyme activity during spray-drying and also on subsequent storage of the dried product. Mumenthaler et al.'s (1994) study of the spray-drying of recombinant human growth hormone (rhGH) and tissue-type plasminogen activator (tPA) provided evidence that denaturation of protein at the air/liquid-interface of droplets in the atomized spray is the major factor promoting protein degradation during spray-drying. It was also shown that addition of surfactant to the spray solution reduced the extent of protein aggregation.

Patton et al.'s (1995) patent covering systemic delivery of insulin by dry powder inhalation specifies spray-drying as being the preferred method of production to yield a high fine-particle dose. The use of amorphous stabilizers, such as disaccharides or amino acids, to enhance stability of insulin during storage is also covered, although no experimental data are given. One year later, Mehta et al. (1996) patented the spray-drying of recombinant human erythropoietin (rhEPO). Again, the powder was considered to be suitable for pulmonary delivery by virtue of its favorable particle size. The spray-dried rhEPO powders showed the same storage stability independent of the addition of stabilizing excipients such as mannitol or amino acids. Thus for spray-drying, the utility of specific stabilizers excipients might be expected to be protein dependent. However, as has been found for freeze-drying (see Chapter by Carpenter et al. in this volume), the disaccharides sucrose or trehalose should provide a high degree of stabilization to most proteins during spray-drying, assuming that appropriate processing conditions are employed (see below).

In the mid- and late-90's, Maa and colleagues published a series of excellent papers about spray-dried particles containing rhGH or other pharmaceutical proteins. The size and morphology of spray-dried protein powders as a function of spray-drying conditions and formulations were presented (Maa et al., 1997). Of especial interest was the finding that the addition of surfactant to the spray solution ameliorated the highly-wrinkled appearance of protein-containing particles. Subsequent experiments showed that aggregation of rhGH during atomization was induced primarily by interactions of the protein with air/liquid-interface (Maa et al., 1997b). In the presence of surfactant, this aggregation could be delayed or prevented. This was unequivocal evidence to show that protein aggregation during spray-drying is more a consequence of adsorption at the air/liquid-interface of the atomized spray than of either shearing or temperature stresses. Surfactant reduced the formation of insoluble protein aggregates even in pure spray-dried rhGH (Maa et al., 1998b). Zinc ions were found to suppress the formation of soluble rhGH aggregates. In a subsequent publication (Maa et al., 1998a) it was shown that the water content of spray-dried, recombinant humanized anti-IgE monoclonal antibody (rhuMAbE25)/mannitol was determined by the humidity of the environment where the powder was manufactured, processed and stored.

Franks' (1991) immobilization hypothesis in the glassy state predicts that the mass-ratio of protein to glassy-carrier will be a vital factor affecting protein stability. Aggregation of rhuMAbE25 during spray-drying was minimized at a trehalose/protein mass-ratio greater than approximately 1 : 1 (Andya et al., 1999). Similar behavior was seen with lactose and, to a lesser extent, with mannitol. It was also demonstrated that maintenance of at least a fraction of mannitol in the amorphous state is important for protein stability in spray-dried powders (Costantino et al., 1998), as is already known for freeze-dried proteins (Pikal, 1994).

Chan and Gonder (1998) published a single study on spray-dried recombinant human deoxyribonuclease (rhDNase) in lactose. They showed that moisture uptake into the product led to lactose crystallization at relative humidities >17%. Broadhead et al. (1996) demonstrated that such moisture uptake is detrimental to the fine-particle dose of spray-dried powders. In 1999, Adler and Lee (1999) published a systematic study of the effects of spray-drying temperature on the residual water content, T_g and the stability of lactate dehydrogenase (LDH) in trehalose. These authors closely examined differences between process parameters and storage stability of the protein. They also used electron spectroscopy for chemical analysis (ESCA) to demonstrate that a surfactant displaced LDH from the surface of the spray-dried particles, this being the reason for the stabilizing action of surfactants during spray-drying. In the same year, Tzannis and Prestrelski (1999a,b) determined the concentration dependence of stabilization of spray-dried trypsinogen by sucrose. As previously found by Andya et al. with trehalose (1999), increasing sugar concentration produced better protein stabilization. At a mass-ratio of sugar/protein = 1 : 1 maximum structural and activity preservation of the trypsinogen was achieved. At higher sucrose concentrations, however, destabilization of the protein was evident, owing possibly to phase separation of the sucrose and trypsinogen in the spray-dried formulation.

Spray-drying by its nature can yield valuable information about protein/carbohydrate interactions in both the liquid and the solid state. Little is known about phase separation of proteins and carbohydrates in concentrated solutions; preferential exclusion (Arakawa and Timasheff, 1992) points the way to further knowledge. Droplets require only a few milliseconds during spray-drying before solid precipitates and prevents further molecular motion (Adler and Lee, 1999). Therefore, this is the time scale on which protein-excipient phase separation and the adsorption of proteins, carbohydrates or surfactants to the air/water-interface can occur. The further study of these phenomena in the liquid and solid state is both of intrinsic scientific merit and valuable practical application for pulmonary protein delivery.

THE PRACTICE OF SPRAY-DRYING PROTEINS

Type of Equipment

Published studies on the spray-drying of proteins use either recombinant therapeutic proteins such as rhGH (Mumenthaler, 1994; Maa et al., 1998b), rhEPO (Mehta et al., 1996) recombinant human granulocyte-colony stimulating factor (rhG-CSF) (Niven et al., 1994) and tPA (Mumenthaler et al., 1994), or enzymes as "model proteins", e.g., trypsinogen (Tzannis and Prestrelski, 1999) and LDH (Adler and Lee, 1999). For research and development purposes the latter have the advantages of ready availability and the possibility to measure their biological activity accurately and directly. Both types of protein are, however, usually expensive and/or available only in small quantities. For this reason the preferred equipment in both industry and universities is a small, laboratory-scale spray-dryer. Figure 1 illustrates the Büchi model 190 co-current

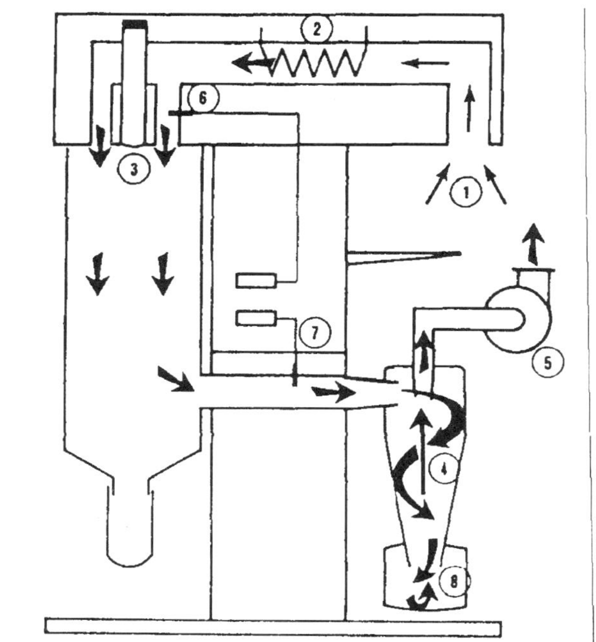

Figure 1. Schematic of Büchi laboratory spray dryer. (1) air inlet; (2) inlet-air heating element; (3) two-fluid nozzle; (4) cyclone separator; (5) ventilator fan; (6) thermocouple to measure T_{inlet}; (7) thermocouple to measure T_{outlet}; (8) product collector. Reproduced with permission of the manufacturer.

spray-dryer, which has been very extensively (almost exclusively) cited in the literature. The unit is fitted with a water-cooled, two-fluid nozzle, usually of 0.7 mm internal diameter. An inlet-air filter is available to prevent particulate contamination of the product. Alternatively, a self-constructed 0.22 μm-membrane filter assembly can be used (Maa et al., 1997). Two thermocouples measure the inlet and outlet air temperatures, T_{inlet} and T_{outlet}, respectively. Thermocouples have also been placed within the drying chamber to enable measurement of drying-air temperature at different distances from the nozzle (Maa et al., 1997). The inside glass wall of the cyclone separator is coated, which may help to improve the yields of sticky sugars. The outlet-air filter protects the vane of the aspirator from particulate contamination.

With this unit a liquid volume as small as 5 ml can be spray-dried, which is 100 times smaller than that required for the Niro Mobile Minor. Total solids' concentrations of up to approx. 20% can be spray-dried. The independent process variables are T_{inlet}, liquid feed rate, atomizing-air volumetric flow rate, and drying-air volumetric flow rate. The single dependent process variable is T_{outlet}.

Spray-Drying Conditions

The selection of suitable spray-drying conditions ('process variables') has a major impact on product properties, viz residual water content, T_g, particle size and morphology, and the extent of protein aggregation and/or inactivation.

Spray-Drying Temperature (T_{inlet}, T_{outlet}). These are the most important process variables. The drying of a spray droplet containing dissolved protein/disaccharide in water can be divided into two phases (Masters, 1991). In the initial *constant rate period*, the water concentration at the droplet surface, and hence the rate of evaporation, is constant. This period lasts until a solid phase begins to form at the droplet surface, initiating the subsequent *falling rate period*. The rate of evaporation now decreases, and the particle size is assumed not to change further. The durations of both drying periods can be calculated from the equations given by Masters (1991). The two-fluid nozzle fitted to the Büchi laboratory spray-dryer produces water droplets of diameters of <10 μm (Maa et al., 1998b), although the droplet size depends to some extent on liquid feed rate and atomizing-air volumetric flow rate (see below). At a high $T_{inlet} = 150°$ C the droplet drying times for a 10% w/w trehalose solution in water in the constant rate and falling rate periods are just 0.4 and 1.1 ms, respectively (Adler and Lee, 1999). It is these short drying times that make spray-drying attractive for thermolabile proteins. Additionally, the temperature at the droplet surface during the constant rate period is maintained at the wet bulb temperature (T_{wb}), which lies substantially below T_{inlet} (e.g., $T_{wb} = 55°$ C at $T_{inlet} = 150°$ C in the above example).

Whereas T_{inlet} is an independent process variable, T_{outlet} is dependent on the liquid feed rate, the atomizing-air volumetric flow rate and the drying-air volumetric flow rate. Maa et al. (1997) demonstrated for the Büchi spray-dryer how T_{outlet} increased linearly with either increasing T_{inlet} or drying-air volumetric flow rate. T_{outlet} decreased linearly with either increasing liquid feed rate or atomizing-air volumetric flow rate. These relations are reproduced in Figure 2 for reference.

Furthermore, the air temperature within the drying chamber corresponded much more closely to T_{outlet} than T_{inlet}, even just 5 cm below the nozzle. The authors concluded that T_{outlet} is therefore more representative than T_{inlet} as being that temperature in the immediate vicinity of the drying droplet. With the Büchi apparatus, however, only T_{inlet} is specified by the operator; T_{outlet} can only be measured during spray-drying.

The maximum T_{inlet} achievable with the Büchi spray-dryer is approx. 200° C. Most studies of the spray-drying of proteins have, however, been performed at much lower T_{inlet} because of concerns about protein thermolability and process cost.

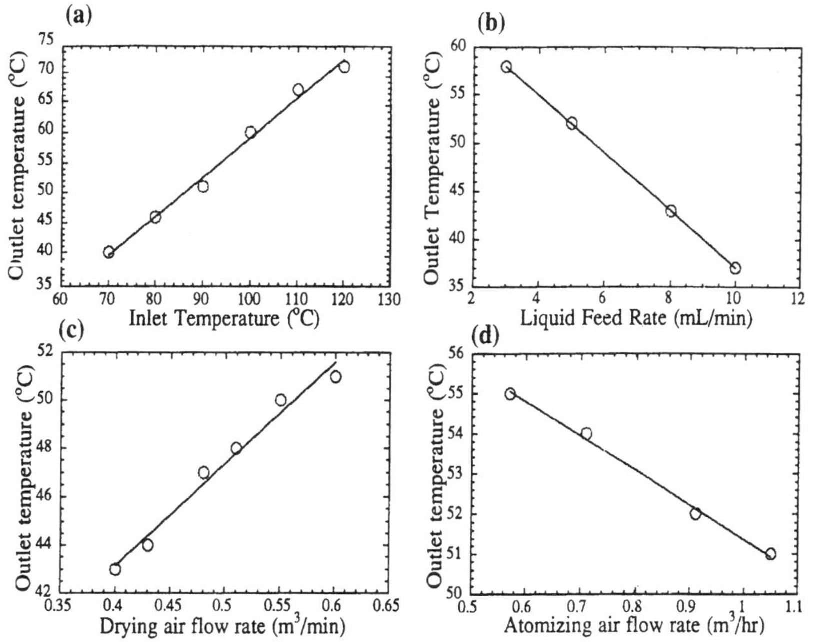

Figure 2. Measured dependencies of T_{outlet} in the Büchi laboratory spray-dryer on (a) T_{inlet}; (b) v_{lf}; (c) v_{da}; (d) v_{aa}. Standard spray-drying conditions used were: $T_{inlet} = 90°C$, $v_{lf} = 5$ ml/min, $v_{da} = 900$ L/h, $v_{aa} = 600$ L/min. Figure reproduced from Maa et al. (1997), with permission.

Increasing T_{inlet} from 50–120°C reduced the residual water content of spray-dried oxyhemoglobin/sucrose (Figure 3), there being no observed dependence on T_{outlet} (Labrude et al., 1989). Even at the highest T_{inlet} used (120°C) the residual water content of 2.3% obtained would give an estimated T_g for sucrose of only approximately 40° C (although this was not reported in the paper). The higher T_{inlet} was, however, detrimental to the process stability of the protein. At T_{inlet} = 120°C only 90% of the original oxyhemoglobin sprayed was detected in the sucrose particles immediately after spray-drying. Product yields on spray-drying are extremely substance-specific and are always a cause for concern. Increasing T_{inlet} from 60 to 120° C increased the yield of the oxyhemoglobin/sucrose from 40 to 70%. The higher T_{inlet} also gave better storage stability.

In contrast, Broadhead et al. (1993) claim that T_{inlet} in the range 70 to 140° C had little effect on the residual water content of pure spray-dried β-galactosidase. However, when the protein was spray-dried together with trehalose, a reduction in the residual water content occurred on increasing T_{inlet}. T_{inlet} also had a major influence on the product yield: increasing T_{inlet} improved the yield, although this effect was coupled with that of the liquid feed rate, v_{lf}. At high v_{lf} the yields were low, irrespective of T_{inlet}. This finding stresses the practical importance of interactions between the independent processes variables (Maa et al., 1997). Broadhead et al. (1993) also confirmed that T_{outlet} depended on both T_{inlet} and v_{lf}. The influence of drying-air volumetric flow rate and atomizing-air volumetric flow rate were, however, not considered, and, as shown by Maa et al. (1997), are relatively weak. Change in T_{inlet} from 60 to 140° C increased the inactivation of β-galactosidase measured immediately after spray-drying from 15% to approximately 65%.

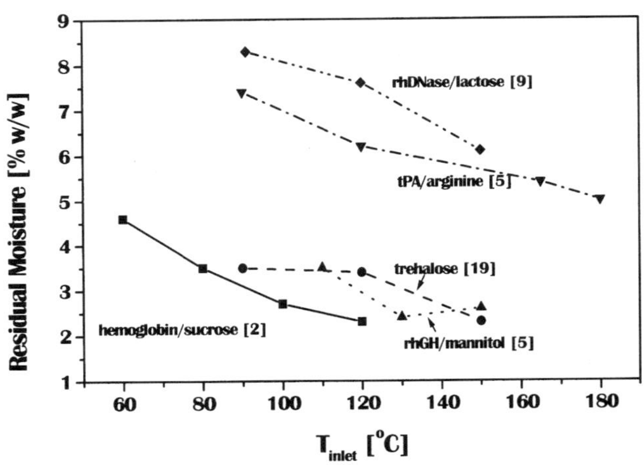

Figure 3. Relation between residual moisture and T_{inlet} for various spray-dried powders taken from the reference cited in the figure. The other independent process parameters (v_{fl}, v_{aa} & v_{da}) were not in all cases the same in the various works. See text for explanations of the abbreviations used.

However, there was a non-linear relationship between T_{outlet} and β-galactosidase inactivation, underlining the complex nature of T_{outlet} as a dependent process variable. In practice it is better to declare the coupled variable T_{inlet}/T_{outlet}. The v_{lf} also influenced protein inactivation because of the coupling of T_{inlet} and v_{lf} to determine T_{outlet}. The authors performed a factorial analysis of their results, which proved particularly valuable in identifying this coupling of T_{inlet} and v_{lf}.

Increasing T_{inlet} reduced the residual water contents of spray-dried rhGH/mannitol and tPA/arginine (Mumenthaler, 1994) (Fig. 3). In the former, however, approx. 25% soluble and insoluble aggregates were identified, irrespective of T_{inlet} between 90 and 150°C and the corresponding T_{wb}'s of 33 to 41° C. Differential scanning calorimetric analysis of aqueous rhGH showed an endothermic transition at approx. 76°C, although treatment in a water bath at 80°C for up to 40 seconds produced only 2% soluble aggregates. tPA showed minimal formation of aggregates at T_{inlet} up to 165° C, although differential scanning calorimetry showed an endothermic transition for the aqueous protein at approx. 70°C. A 26 second exposure of a tPA solution to 80°C also gave only negligible degradation. The authors conclude that a process other than heat stress must play a central role in protein degradation during spray-drying. The other factors contributing to protein aggregation will be discussed in more detail below.

By varying both T_{inlet} and v_{lf} Maa et al. (1997) showed that an increase in T_{outlet} reduced the residual water content of spray-dried rhDNase/lactose (Figure 4). However, there was also a linear relation between T_{inlet} and residual water content (Fig. 3).

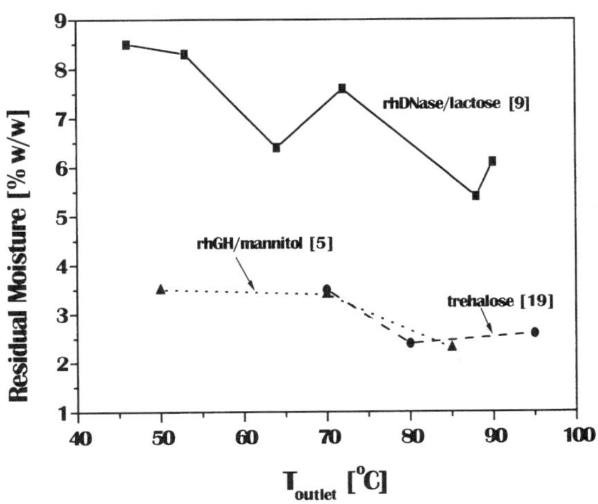

Figure 4. Relation between residual moisture and T_{outlet} for various spray-dried powders taken from the reference cited in the figure. The other independent process parameters (v_{fl}, v_{aa} & v_{da}) were not in all cases the same in the various works.

Dried particle size of rhDNase/lactose (Maa et al., 1997) tended to increase slightly with higher T_{inlet}/T_{outlet}, as also noted by Broadhead et al. (1993) with β-galactosidase. There was a stronger observed dependence of particle shape on T_{outlet}: increase in T_{outlet} caused greater deformation (e.g., dimples and holes) of the dried particles. However, this result needs to be treated with some circumspection, because T_{outlet} was varied for constant T_{inlet} by decreasing v_{lf} , the influence of which on particle shape is unclear. The correlation between T_{inlet} and dried particle shape by constant v_{lf} is, judged visually, not so strong. Adler and Lee (1999) produced spray-dried trehalose/LDH particles in which residual water contents were linearly dependent on both T_{inlet} (Fig. 3) and T_{outlet} (Fig. 4). Increasing T_{inlet}/T_{outlet} also caused a substantial improvement in yield of spray-dried trehalose, being approx. 80% at T_{inlet} = 150° C, a good result with the Büchi spray-drying unit. This yield was achieved at the "high" v_{lf} of 4 ml/min, where Broadhead et al. (1993) found low yields irrespective of drying temperature of β-galactosidase.

T_{inlet}/T_{outlet} influenced the process and storage stabilities of LDH in spray-dried trehalose in quite different manners (Adler and Lee, 1999). At mild T_{inlet}/T_{outlet} = 90° C/60° C a process loss of approximately 11% of LDH catalytic activity occurred, increasing to approximately 25% at T_{inlet}/T_{outlet} = 150° C/95° C (see Fig. 5). The product obtained with T_{inlet}/T_{outlet} = 90°C/60°C showed a subsequent continual decline in LDH activity at all four storage temperatures examined (Fig. 5). The spray-dried LDH/trehalose obtained at T_{inlet}/T_{outlet} = 150° C/95° C showed, however, no loss in LDH activity during storage at temperatures up to room temperature. This systematic study confirmed the early observation of Labrude et al. (1989) that higher T_{inlet} reduced process stability but improved storage stability of spray-dried oxyhemoglobin. The improvement in storage stability at higher T_{inlet}/T_{outlet}, as seen in Figure 5, runs parallel to reduced residual water contents of the powders. The authors found, however, no evident dependence of storage stability at the four temperatures on the T_g of the particles. The complex relation between water content, T_g, and protein stability has, however, yet to be resolved (cf., Hancock and Zografi, 1997; Pikal, 1999).

Apart from irreversible protein inactivation, processes such as spray- or freeze-drying may also induce reversible loss of protein native form (Heller et al., 1999; Tzannis and Prestrelski, 1999a). This necessitates determination of both protein aggregation status and activity when examining stability on drying, clearly a problem with most recombinant proteins.

The influence of T_{inlet}/T_{outlet} on residual water content and protein stability during and after spray-drying is thus a vital factor. There is also an apparent correlation between residual water content of the product and the relative humidity of the inlet-air on the day of a spray-drying experiment, which partially overrides the influence of T_{inlet}/T_{outlet} (Adler and Lee, unpublished observations). This phenomenon had already been observed by Broadhead et al. (1993), who noted that

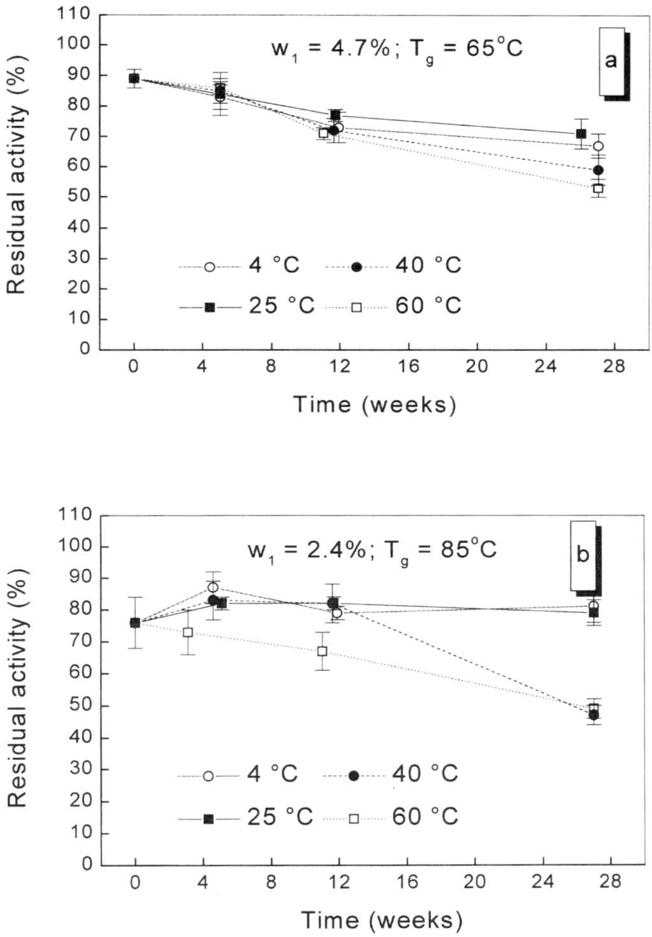

Figure 5. Residual activity of LDH measured immediately after spray-drying at T_{inlet}/T_{outlet} = (a) 90°C/60°C, or (b) 150°C/95°C, and also on subsequent dry-storage at various temperatures. Reproduced from Adler and Lee (1999) with permisssion.

the residual water contents of numerous batches of spray-dried trehalose could be plotted against relative humidity to give a more-or-less linear dependence. In the Büchi unit a linear relationship between the relative humidity of the air in the cyclone separator/receiver and T_{outlet} has been found (Maa et al., 1998a). It follows that the product in the receiver will be influenced by the relative humidity there, although the product will be in non-equilibrium because of the short experimental duration. Knowledge of a powder's moisture sorption isotherm enables estimation of the driest powder that a spray-dryer can produce. For the example of rhuMAbE25/mannitol (80 : 20 by weight), the calculated value was approximately 3–4% water by mass at $T_{outlet} > 80°C$ (Maa et al., 1998a), which necessitates a high T_{inlet} and/or high liquid feed rate. These results stress the utility of using conditioned inlet air even with the laboratory-scale Büchi spray-dryer. This approach could simplify understanding of the relation between T_{inlet}/T_{outlet} and relative water content of the product and decrease batch-to batch variation.

Liquid Feed Rate (v_{lf}) and Atomising Air Volumetric Flow Rate (v_{aa}). v_{lf} is a major influencing factor on T_{outlet}, and combined with v_{aa}, also determines the spray droplet size. As already discussed above (see Fig. 2), increasing v_{lf} at a fixed T_{inlet} will decrease T_{outlet} (Maa et al., 1997). With the Büchi laboratory spray-dryer fitted with the standard two-fluid nozzle, the slowest practical v_{lf} is 1 ml/min. Most reported studies of the spray-drying of proteins using the Büchi use a v_{lf} of approximately 5 ml/min. Some reports quote a v_{lf} of 15 ml/min (Maa et al., 1998a; Andya et al., 1999), but this necessitates a drying-air volumetric flow rate of 1000 L/h in place of the usual 600 L/h. If v_{lf} is too high for the other conditions used, then condensation appears in the drying chamber (Broadhead et al., 1993).

The coupling of v_{lf} with v_{aa} and their influence on spray droplet size is defined by the so-called air/fluid mass ratio, $\mathbf{n}_{a/f}$, which represents the energy available for atomization:

$$\mathbf{n}_{a/f} = \frac{v_{aa}[m^3/h] \cdot \text{air density}[kg/m^3]}{v_{if}[m^3/h] \cdot \text{fluid density}[kg/m^3]} \quad (1)$$

Increase in $\mathbf{n}_{a/f}$ represents more energy and therefore reduces spray droplet size and consequently the size of the dried particles (Masters, 1991). In practice, however, v_{lf} is constrained because of its influence on T_{outlet}. v_{aa} lies between 500 to 1000 L/h. Therefore, the possibility to tailor product particle size via these parameters on the Büchi laboratory spray-dryer is very limited. A value of $\mathbf{n}_{a/f}$ of 3 to 5 will exist, making only moderate variation in spray droplet size possible (Masters, 1991).

The diameter of a solid particle produced by spray-drying, D_s, is dependent on the diameter of the original spray-droplet, D_w, and also the total solids' content of the spray solution, w_2, (Hickey et al., 1994):

$$D_w = 2 \cdot \sqrt[3]{\left((D_S/2)^3 \rho_2\right)/w_2} \quad (2)$$

where ρ_2 is the density of the solid. Decreasing the trehalose concentration in a spray solution from 10 to 2% w/w thus reduced the mean D_s measured by laser diffraction from 3.5 to 2.2 μm, when the solution was spray-dried at $T_{inlet}/T_{outlet} =$

150°C/95°C (Adler and Lee, 1999). However, no correlation was found between total solids' content of the spray solution and D_s for rhDNase/lactose spray-dried at $T_{inlet}/T_{outlet} = 90°C/53°C$ (Maa et al., 1997). In summary, there is only a little room to maneuver in controlling dried particle size using the Büchi spray-dryer. It is fortuitous that the range of process conditions available on the Büchi unit produces dried particles of size suitable for pulmonary delivery. Should larger particles be required, then the Büchi unit is no longer suitable. Instead, a larger spray-dryer, such as the Niro Mobile Minor fitted with a spraying wheel, is appropriate, in which values of $\mathbf{n}_{a/f}$ of 0.5–1.5 can be achieved (Masters, 1991) giving $D_s > 20\,\mu m$ (Adler and Lee, unpublished observation).

Drying-Air Volumetric Flow Rate (v_{da}). This independent process variable can be adjusted to between approx. 100 to 1000 L/min on the Büchi. Increasing v_{da} on the Büchi unit produced an increase in T_{outlet} (Maa et al., 1997) (Fig. 2). In practice, however, this relation is not exploited to control T_{outlet}, since T_{inlet} and v_{lf} have greater effects. The usual value for v_{da} quoted in the literature is 600 L/min on the Büchi unit combined with a v_{lf} of 5 ml/min. At higher v_{lf} (e.g., 15 ml/min) v_{da} needs to be increased to approximately 1000 L/min (Maa et al., 1998a; Andya et al., 1999). Very low v_{lf}, for example 1 ml/min, can apparently be used together with a v_{da} of 500 L/min (Mehta et al., 1996). This process parameter is therefore only of minor importance in controlling product properties.

Influence of Formulation

Although solutions of some pure proteins have been successfully spray-dried, adequate formulation is almost always necessary to ensure adequate protein process and storage stability. Indeed, most of the literature of the last 10 years is concerned with the effects of various excipients on product quality. Adequate formulation is therefore a vital question during the development of a spray-dried protein powder. Much of the knowledge available about formulation of freeze-dried proteins can, however, also be applied to spray-dried systems.

Pure Proteins

A number of authors have attempted to spray-dry pure proteins, for example oxyhemoglobin (Labrude et al., 1989), rhEPO (Mehta et al., 1996), β-galactosidase (Broadhead et al., 1993), trypsinogen (Tzannis and Prestrelski, 1999b), rhGH (Maa et al., 1998b), and rhuMAbE25 (Andya et al., 1999). In all cases, even at low T_{inlet} there is substantial aggregation or inactivation of the protein when measured in samples rehydrated immediately after spray-drying ("process" instability). The following three examples illustrate all of the relevant factors and product properties.

On spray-drying a 2 mg/ml aqueous solution of rhGH at T_{inlet}/T_{outlet} = 90° C/53° C (v_{da} = 600 L/h, v_{aa} = 900 L/h, v_{lf} = 5 mL/min) extremely wrinkled particles were obtained, with a diameter 3.0 μm, a residual content of 11% and an approximately 55% yield (Maa et al., 1998b). This morphology was also found with spray-dried bovine serum albumin (bSA), but not with rhDNase, which had smooth, round particles (Maa et al., 1997). There was no correlation between surface tension of the protein solutions and resulting dried particle morphology. Most likely differing porosity, flexibility and mechanical strength of the dry crust formed in the falling rate period of droplet drying are responsible for these differing morphologies. Pure rhGH suffered extensive damage during spray-drying, producing a very cloudy solution after reconstitution containing 32% insoluble and 7% soluble aggregates (Maa et al., 1998b). Clearly, formulation measures are necessary to reduce this damage during spray-drying.

On spray-drying rhuMAbE25 under the same conditions used for rhGH quoted above, the pure protein particles had smooth surfaces but showed large indentations (Maa et al., 1997), illustrating again the unpredictable variety of particle forms obtained on spray-drying of pure proteins. The dried particle median diameter was 3.2 μm. At T_{inlet}/T_{outlet} = 105° C/55° C the rhuMAbE25-particles contained 5.6% soluble aggregates, mostly being dimers. This protein is evidently less sensitive to spray-drying stresses than rhGH. Over 1 year of storage the spray-dried pure protein showed a continued decrease in monomer levels ("storage" instability), with first order rate constants of 0.5×10^{-3} day^{-1} and 2.2×10^{-3} day^{-1} at 5°C and 30°C, respectively. Again, formulation with the appropriate stabilizing excipients is necessary to improve both the process and storage stability of this protein.

A study of trypsinogen (Tzannis and Prestrelski, 1999a) is of particular value, because it throws light on protein structure within the solid state produced by spray-drying. On spray-drying the pure protein at T_{inlet}/T_{outlet} = 120° C/90° C and v_{lf} = 5 ml/min the dried particle mean diameter was 3.1 μm with a residual water content of 2.2%. Infrared spectroscopic analysis indicated that the dried protein had non-native secondary structure. Upon rehydration trypsinogen had an approximately 15% decrease in enzymatic activity, and its enthalpy of unfolding (determined after rehydration by differential scanning calorimetry) decreased by the same percentage. Yet it lost only 8.4% of its monomer content, this being paralleled by an increase in dimer content of 8.3%. The authors propose that the difference between 15% activity loss and 8.4% aggregation can be assigned to irreversibly unfolded protein monomers. Furthermore, they note that both formation of aggregates and irreversibly unfolded monomers can be attributed to protein unfolding during spray-drying. Upon rehydration there is kinetic competition between refolding and aggregation, with a small fraction of unfolded molecules remaining as unfolded monomers. Again, formulation must be exploited to prevent these detrimental effects to protein structure during spray-drying.

Formulated Systems

The use of carbohydrates to stabilize spray-dried proteins was adopted from the freeze-drying literature. Many disaccharides stabilize proteins in solution against aggregation/inactivation during thermal stress, including that arising during spray-drying, and freezing ("cryoprotectant" action), most likely via a preferential exclusion mechanism (Pikal, 1994; Carpenter et al., 1994). They also form amorphous phases that are effective in stabilizing proteins during drying and also on subsequent storage (Franks, 1990, 1998). Sucrose, trehalose, maltose and lactose are effective "lyoprotectants" for a number of proteins, whereas crystallizing mannitol offers less protecting ability. Sucrose and trehalose have the additional advantage of being non-reducing sugars and avoid the Maillard reaction with proteins. The "glass immobilization" hypothesis propounded by Franks (1990), postulates that a protein is physically immobilized within the amorphous glass existing between the sugar's glass transition temperature, T_g, and its Kauzman temperature, K_z. The "water replacement" hypothesis promoted by Carpenter et al. (1994) claims that the carbohydrate molecules form hydrogen bonds with the protein thereby serving as a substitute for water and inhibiting dehydration-induced protein unfolding. Numerous studies with freeze-dried proteins (reviewed in the Chapter by Carpenter et al. in this volume) and a single study with spray-dried trypsinogen (Tzannis and Prestrelski, 1999b) have documented that effective protection of protein structure during dehydration depends of the presence of a carbohydrate that both remains amorphous and hydrogen bonds to the dried protein.

When formulating a protein for spray-drying it is vital to select a suitable stabilizer and the most efficacious mass ratio of stabilizer/protein. At a mass ratio sucrose/oxyhemoglobin of 0.66 : 1 only 1.5% methhemoglobin was detected immediately after spray-drying oxyhemoglobin, compared with 50% for the pure protein (Labrude et al., 1989). After 10 weeks of storage at 4° C, however, the amount of methhemoglobin rose to 7%. The presence of sucrose, arginine-chloride or trehalose in spray-dried β-galactosidase completely inhibited loss of protein activity during spray-drying (Broadhead et al., 1993). The trehalose formulation gave especially good results and showed no activity loss during one year of storage. Mannitol gave only a slight stabilizing effect on both the process and storage stability of the protein. These authors did not consider that trehalose forms an amorphous solid whereas mannitol is crystalline, and offered no explanation of their findings. However, they did observe that those excipients providing the greatest process stability also afford the best storage stability. Additionally, they report striking differences in particle morphology: the smooth-surfaced, donut-shaped particles of pure β-galactosidase contrast starkly with the pitted surfaces of sucrose particles or the highly-wrinkled surfaces of trehalose particles. This wide-ranging study thus laid the foundations for the use of formulation measures to improve protein stability during spray-drying.

Neither mannitol nor glycine was found to improve the process or storage stability of spray-dried rhEPO (Mehta et al., 1996). Both of these excipients crystallize, on spray-drying at the concentrations quoted in this reference, and a stabilizing action on the protein would therefore not be expected. Similarly, Patton et al.'s (1995) patent describes the spray-drying of insulin with either mannitol (certainly crystalline) or raffinose; the latter is amorphous after freeze-drying (Kajiwara et al., 1999). Byron et al. (1996) spray-dried sucrose, lactose and trehalose using a Yamato ADL-51 mini-spray-dryer, neglecting to state the process conditions. Wide-angle X-ray diffraction (WAXS) measurements showed amorphous particles with high fine-particle doses (>50%), which rapidly crystallized on storage at relative humidities ≥52%. In contrast, spray-dried mannitol was fully crystalline.

Details of the morphology of spray-dried protein/sugar-mixtures are given by Maa et al. (1997). On formulating rhDNase with increasing amounts of lactose, the spray-dried particles of pure rhDNase lost their round smoothness and became pitted. When formulated with increasing amounts of mannitol, an apparently crystalline surface was seen, which resembled the pattern of composite, elongated crystals seen in the surface of spray-dried, pure mannitol (Chew and Chan, 1999). WAXS measurements confirmed that when the concentration of mannitol is ≥ approx. 30% of the total solids it is susceptible to crystallization during spray-drying. Consequently, the process stability of spray-dried rhuMAbE25/mannitol was best at approx. 30% mannitol fraction, decreasing dramatically at higher fractions because of mannitol crystallization (Costantino, 1998). To avoid this detrimental effect, sodium phosphate was added to the spray-dried solution to inhibit mannitol crystallization, thereby lowering the degree of protein aggregation during and after the process. Formulations of rhuMAbE25/trehalose were reported to give highly agglomerated particles, attributed to trehalose's "sticky" nature (Maa et al., 1997). As we shall see below, however, this problem can be avoided by the use of higher T_{inlet} than that of 90° C used in the cited work.

The best sugars for stabilizing sensitive proteins during and after spray-drying are sucrose and trehalose. But which mass ratio of sugar : protein gives the best process and storage stability of the protein? The formation of soluble aggregates by rhuMAbE25 on spray-drying at T_{inlet}/T_{outlet} = 105° C/55° C was reduced from 5.6% for the pure protein to <1% when formulated with either trehalose or lactose (a reducing sugar!) at a sugar : protein mass ratio ≥0.2 : 1 (Andya et al.,

1999). It is unclear, however, if a mass ratio of only 20% sugar can build an amorphous matrix in which the protein at 80% mass fraction is embedded. Indeed, a higher sugar : protein mass ratio of 1 : 1 was necessary to assure maximum storage stability up to 1 year at 30°C. The storage stability results could therefore be explained by the "glass immobilization" hypothesis, even though the authors reject the importance of this idea. The authors similarly reject the importance of the "water replacement" hypothesis for their results, since they detected no qualitative change in the infrared spectrum of the protein after spray-drying or storage. A more exact interpretation of the results in this paper is, however, hindered by a lack of thermo-analytical data. Knowledge of the existence of a glass transition, for example, could have illuminated the relation between sugar : protein mass ratio and glassy-matrix formation. Nonetheless, this paper also contains the interesting result that the process and storage stability of rhuMAbE25 in mannitol was improved with up to approx. 40% mannitol, where the mannitol remained amorphous. Higher mannitol concentrations caused crystallization and hence less protein stabilisation. Altogether, mannitol was not as efficient a stabilizer as the other sugars, even when amorphous.

A sucrose/protein mass ratio of 0.25 : 1 reduced the inactivation of trypsinogen on spray-drying at T_{inlet}/T_{outlet} = 120° C/85° C from 15% for the pure protein to 6%, falling further to 0% the at optimum sucrose : protein mass ratio of 1 : 1 (Tzannis and Prestrelski, 1999a). An infrared spectroscopic analysis of the dried protein indicated increasing recovery of native secondary structure within this range of sucrose/protein mass ratio. However, further increase to >1 : 1 sucrose : protein mass ratio led to decreased recovery of activity after rehydration, and the infrared pectroscopic data indicated a reduction of native structure in the dried solid. Similarly, differential scanning calorimetry showed a decrease in the enthalpy of unfolding, which correlated with the incurred activity losses. It appears therefore that at sucrose/protein mass ratios >1 : 1 some irreversible structural destabilization of the trypsinogen in the spray-dried sucrose occurred. The authors suggest this is a result of formation of protein-rich and sucrose-rich phases in the spray-dried solid as a consequence of preferential exclusion of sucrose from the protein surface. In addition to reducing the capacity of the sugar to interact favorably with the protein in the dried solid, phase separation creates new interfaces that could promote protein unfolding (Heller et al., 1999). These ideas are of considerable scientific and practical interest, indeed also for freeze-dried systems (Heller et al., 1999).

Use of Added Surface Active Substances

The addition of surface-active substances to spray solutions in order to reduce the process instability of proteins is worthy of special consideration.

Proteins are sensitive to damage at interfaces. The adsorption process of a protein at a mobile interface (air/liquid or liquid/liquid) comprises molecular diffusion to the interfacial region followed by adsorption to the interface and subsequent, long-lasting molecular reorientation (Tripp et al., 1995). This can be accompanied by protein unfolding, leading to aggregation and possibly inactivation. Part of the measured inactivation of proteins during freeze-drying is thought to be a result of adsorption at the (immobile) water/ice-interface (Chang et al., 1996). The situation with spray-drying of a protein solution is more critical, since atomization produces an extremely rapid and very large expansion of the liquid/air-interface. Does protein adsorption at the expanding liquid/air-interface during atomization offer an explanation for observed protein destabilization during spray-drying? Can surface-active substances be employed to improve protein stability?

Broadhead et al. (1993) added 0.05% polysorbate 80 to a β-galactosidase spray solution and observed no significant effect on residual protein activity in samples rehydrated immediately after spray-drying. Because the surfactant reduced the yield substantially, these authors rejected its use as a stabilizer.

An unequivocal indication of the utility of surfactants is found in Mumenthaler et al. (1994). A solution of hGH was atomized into a glass beaker using a two-fluid nozzle at room temperature, resulting in increased formation of insoluble and soluble aggregates with greater v_{aa}. At the corresponding v_{aa}, the liquid sample contained the same amount of insoluble aggregates and two-thirds of the amount of soluble aggregates found in a spray-dried protein sample. This results strongly suggests that during spray-drying the formation of aggregates could be induced by liquid/air-interfacial adsorption and unfolding. The addition of 0.1% polysorbate 20 to the spray solution ameliorated the formation of both insoluble and soluble rGH aggregates during atomization and also spray-drying. The authors suggested for the first time in the pharmaceutical literature that the surfactant displaces protein molecules at the air/liquid-interface and hereby reduces exposure of the protein to aggregation. This intuitively attractive hypothesis has since found wide acceptance.

The various stresses experienced by a protein molecule in aqueous solution during spray-drying are illustrated schematically in Figure 6. The most important published study of the relative importance of these stresses is from Maa and Hsu (1997). Using two ingenious systems it was possible to generate high shear in an aqueous solution either in the presence or absence of an air/liquid-interface. High shear combined with an air/liquid-interface led to formation of non-covalent aggregates of rhGH. The extent of aggregation increased with higher protein concentration or larger air/liquid-interfacial area. In the absence of an air/liquid-interface, however, rhGH remained intact even at shear forces two orders of magnitude higher than those that caused extensive aggregation in the presence of a liquid/air-interface. This result illustrates the critical role played by the air/liquid-interface

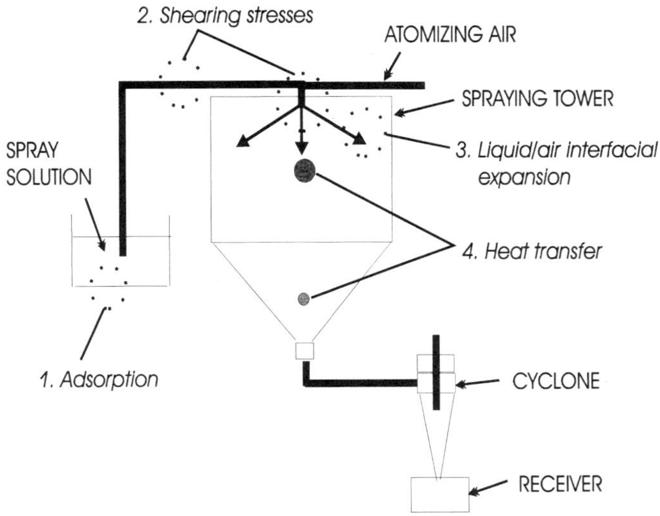

Figure 6. Schematic representation of the possible stresses experienced by a protein during spray-drying. See text for explanations. The current state of knowledge indicates that protein adsorption at the large liquid/air interface created during atomizing of the spray solution is mainly responsible for the protein aggregation/inactivation measured on spray-drying ("process instability" of the protein).

during protein aggregation in solution, at least for rhGH. The presence of 0.1 to 1 mg/ml of either polysorbate 20 or Pluronic F88 greatly reduced the extent of aggregation, with Pluronic F88 being the more effective stabilizer.

The relevance of these results for spray-drying of rhGH was demonstrated in a second paper (Maa et al., 1998b). As previously found for β-galactosidase (Broadhead et al., 1993) and rhGH (Mumenthaler et al., 1994), atomization of an excipient-free rhGH aqueous solution into a beaker ($v_{lf} = 5$ ml/min, $v_{aa} = 900$ L/h) produced large amounts of both insoluble and soluble aggregates. Increasing amounts of polysorbate 20 in the spray solutions reduced the extent of aggregation, giving the best protection at ≥0.05% w/w surfactant. Beyond this "critical polysorbate concentration" the surfactant did not further improve protein stability. By atomizing at different v_{aa}'s to give droplets of different diameter, a linear relationship between the "critical polysorbate concentration" required to give best protection and the total air/liquid-interfacial area was found. Spray-drying the rhGH solution containing 0.05% w/w polysorbate 20 ($T_{inlet}/T_{outlet} = 90°$ C/53° C) produced only 0.5% insoluble aggregates, instead of the 32% aggregated protein noted without surfactant. Taken together, these results show that exposure to the large air/liquid-interface created during atomization is the dominant factor in rhGH aggregation on spray-drying. The authors conclude that the effect of drying stress following atomization "was not significant" (Maa et al., 1998b).

An additional finding was that formation of soluble rhGH aggregates on spray-drying could be reduced by adding Zn^{2+} to the spray solution (Maa et al., 1998a). Combination of surfactant and Zn^{2+} produced an aggregate-free spray-dried rhGH powder, without the necessity of adding a sugar as stabilizer. Presumably a sugar would still be required to guarantee the storage stability of the protein, a point worth investigating. Incidentally, the surfactant had no detrimental effect on either product yield or particle size, although residual water content was lower.

Surfactant also affects the morphology of spray-dried particles. The highly wrinkled morphology of pure spray-dried rhGH particles became much smoother in the presence of polysorbate 20 and/or Zn^{2+} (Maa et al., 1998b). The effect of surfactant on protein/sugar particles is even more marked. For example, the addition of 0.5% w/w polysorbate 20 to spray-dried lactose : rhuMAbE25 (3 : 2) changed the donut-shape to perfect sphericity (Maa et al., 1997).

Pure spray-dried trehalose particles were found to be round with smooth surfaces (Fig. 7) (Adler and Lee, 1999). When loaded with 5% w/w LDH, the particles became wrinkled and deformed (Figure 7). The presence of 0.1% w/w polysorbate in the spray solution returned particle shape and morphology to that seen with pure trehalose (19). Surfactant is thought to reduce internal motions and surface turbulence of droplets uring drying and thereby alter the balance of viscous-to-surface forces governing particle shape (Frey and King, 1996).

Electron spectroscopy for chemical analysis (ESCA) has recently been exploited to examine the surface composition of spray-dried particles containing protein (Adler and Lee, 1999; Fäldt and Bergenstrahl, 1994; Millqvist-Fureby et al., 1999; Adler et al., 2000). The outer surface (approx. 10 nm (Fäldt and Bergenstrahl, 1994)) of spray-dried trehalose/LDH particles was found to contain 7 atomic % N (Adler and Lee, 1999), a value some ten-times greater than that expected for a homogeneous distribution of the LDH within the particles. The drying time of the spray droplets was calculated to be approx. 0.4 ms up to the critical point. Static surface tension measurements showed that sufficient LDH could adsorb to the water/air-interface during this time to explain the measured surface excess in the solid particles. The addition of 0.1% polysorbate 80 to the spray solution reduced the surface concentration of atomic N in the spray-dried trehalose/LDH particles to zero, proving surface exclusion of the protein. Essentially the same result was published for polysorbate 80 in spray-dried sugar/trypsin particles (Millqvist-Fureby et al., 1999). A more thorough study of the system of bovine serum albumin (BSA) and trehalose (Adler et al., 2000) showed that the protein was excluded from the surface of spray-dried solid particles in the presence of either polysorbate or sodium lauryl sulfate (SDS) in a concentration-dependent fashion. The surface concentration of the BSA

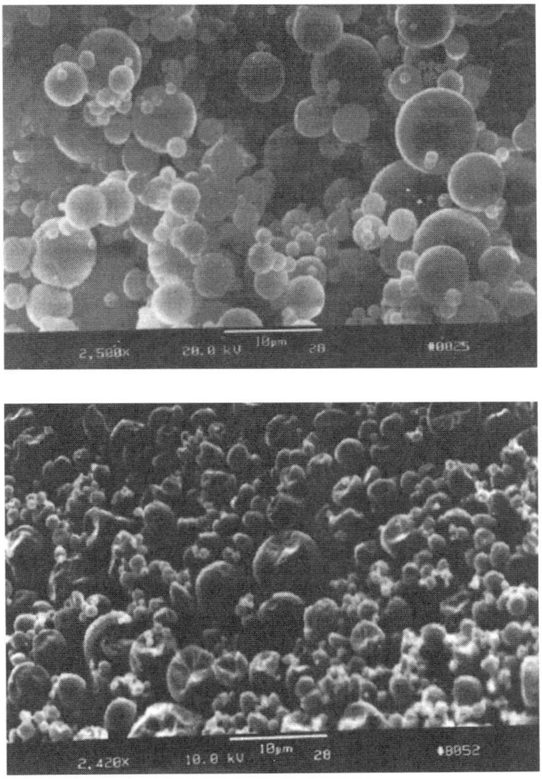

Figure 7. Scanning electron micrographs of pure spray-dried trehalose (left) and spray dried trehalose loaded with 5% w/w LDH (right). Reproduced from Adler & Lee (19) with permission.

decreased, while that of the SDS increased, indicating competitive adsorption of protein and surfactant to the liquid/air interface of the spray droplets up to the critical point. Molar ratios of polysorbate 80 : BSA of ≥1 : 10 or of SDS/bSA of ≥1 : 20 completely excluded the protein from the surface of the spray-dried particles. This progressive exclusion of the protein from the solid surface by the surfactant was accompanied by a parallel alteration of the wrinkled appearance of the surfactant-free particles to smooth spheres.

Although many questions about surfactant/protein/sugar interactions during droplet drying are still unanswered, the practical utility of the use of surfactants to protect proteins during the spray-drying process is unequivocal. Indeed, other surface-active substances not usually considered to be "classical" surfactants also show a stabilizing effect. The presence of 1% w/w hydroxypropyl-β-cyclodextrin in a spray solution reduced the inactivation of β-galactosidase spray-dried at T_{inlet}/T_{outlet} = 190°C/61°C from 25% to <2% (v_{lf} = 6 ml/min; v_{aa} = 750 L/h; v_{da} = 500 L/min) (Branchu et al., 1999). One should not lose sight of the cost, however. Possible effects of the surfactant on the deposition kinetics of protein-loaded spray-dried particles also need to be examined. Surfactant-containing particles appear to be "stickier" than surfactant-free particles, causing them to be retained in higher levels of a cascade impactor (Adler and Lee, unpublished observations).

CONCLUDING REMARKS

The utility of spray-drying to produce protein-loaded particles suitable for pulmonary delivery is evident. The extent of protein aggregation and/or inactivation during and after spray-drying can be minimized by means of formulation measures, such as the inclusion of surfactants and stabilizing sugars. Provided suitable spray-drying conditions can be identified, this process is attractive as a method to stabilize proteins. As with freeze-drying, however, a certain amount of experience is invaluable to enable the successful development of a spray-dried product for a protein drug. This necessitates back-up from the equipment manufacture as well as knowledge of the pharmaceutics of the process, which I hope is given in introductory form in this chapter.

REFERENCES

Adler, M., and Lee, G., 1999. Stability and surface activity of lactate dehydrogenase in spray-dried trehalose. *J. Pharm. Sci.* 88:199.

Adler, M., Unger, M., and Lee, G., 2000. The surface composition of spray-dried trehalose/bovine serum albumin/surfactant particle. *Pharm. Res.* 17:863.

Andya, J., Maa, Y., Costantino, H., Nguyen, P., Dasovich, N., Sweeney, T., Hsu, C., and Shire, S., 1999. The effect of formulation excipients on protein stability and aerosol performance of spray-dried powders of a recombinant humanized anti-IgE monoclonal antibody. *Pharm. Res.* 16:350.

Arakawa, T., and Timasheff, S., 1982. Stabilization of protein structures by sugars. *Biochemistry* 21:6536.

Branchu, S., Forbes, R., York, P., Petren, S., Nyquist, H., and Camber, O., 1999. Hydroxypropyl-β-cyclodextrin inhibits spray-drying-induced inactivation of β-galactosidase. *J. Pharm. Sci.* 88:905.

Broadhead, J., Rouan, S., Hau, I., and Rhodes, C., 1993. The effect of process and formulation variables on the properties of spray-dried β-galactosidase. *J. Pharm. Pharmacol.* 46:458.

Broadhead, J., Rouan, S., and Rhodes, C., 1992. The spray drying of pharmaceuticals. *Drug Devel. Ind. Pharm.* 18:1169.

Broadhead, J., Rouan, S., and Rhodes, C., 1996. The deposition of spray-dried β-galactosidase from dry powder inhaler devices. *Drug Devel. Ind. Pharm.* 22:813.

Byron, P., Naini, V., and Phillips, E., 1996. Drug carrier selection—important physio-chemical characteristics, in: *Respiratory Drug Delivery V*, P. Byron, P. et al. eds., Interpharm. Press, Buffalo Grove, IL.

Carpenter, J., Prestrelski, S., Anchorduguy, S., and Arakawa, T., 1994. Interactions of stabilisers with proteins during freezing and drying, in: *Formulation and Delivery of Proteins*, J.L. Cleland and R. Langer, eds., American Chemical Society, Washington, DC.

Chan, H., and Gonda, I., 1998. Solid state characterisation of spray-dried powders of recombinant human deoxyribonuclease. *J. Pharm. Sci.* 87:647.

Chang, B., Kendrick, B., and Carpenter, J., 1996. Surface-induced denaturation of proteins during freezing and its inhibition by surfactants. *J. Pharm. Sci.* 85:1325.

Chew, N., and Chan, H., 1999. Influence of particle size, air flow and inhaler device on the dispersion of mannitol powders as aerosols. *Pharm. Res.* 16:1098.

Costantino, H., Andya, J., Nguyen, P., Dasovich, N., Sweeney, T., Shire, S., Hsu, C., and Maa, Y., 1998. Effect of mannitol crystallisation on the stability and aerosol performance of a spray-dried pharmaceutical protein, recombinant humanized anti-IgE monoclonal antibody. *J. Pharm. Sci.* 87:1406.

Fäldt, P., and Bergenstrahl, B., 1994. The surface composition of spray-dried protein-lactose powders. *Coll. Surf.* A90:183.

Franks, F., Freeze-drying of bioproducts: putting principles into practice. *Eur. J. Pharm. Biopharm.* 45:221.

Franks, F., 1990. Freeze-drying: from empiricism to predictability. *Cryo-Letters* 11:93.

Frey, D., and King, C., 1986. Effect of surfactants on mass transfer during spray drying. *AICHEJ* 32:437.

Hancock, B., and Zografi, G., 1997. Characteristics and significance of the amorphous state in pharmaceutical systems. *J. Pharm. Sci.* 86:1.

Heller, M., Carpenter, J., and Randolph, T., 1999. Protein formulation and lyophilization cycle design: prevention of damage due to freeze-concentration induced phase separation. *Biotechnol. Bioeng.* 63:167.

Hickey, A., Concessio, N., Ort, M., and Platz, R., 1994. Factors influencing the dispersion of dry powders as aerosols. *Pharm. Technol.* August:58.

Hiemenz, P., 1986. *Principles of Colloid and Surface Chemistry*, Marcel Dekker, New York.

Kajiwara, K., Franks, F., Echlin, P., and Greer, A., 1999. Structural and dynamic properties of crystalline and amorphous phases in raffinose-water mixtures. *Pharm. Res.* 16:1441.

Labrude, P., Rasolomanana, M., Vigneron, C., Thirion, C., and Chaillot, B., 1989. Protective effect of sucrose on spray drying of oxyhemoglobin. *J. Pharm. Sci.* 78:223.

Maa, Y., Costantino, H., Nguyen, P., and Hsu, C., 1997. The effect of operating and formulation variables on the morphology of spray-dried protein particles. *Pharm. Devel. Technol.* 2:213.

Maa, Y., and Hsu, C., 1997. Protein denaturation by combined effect of shear and air-liquid interface. *Biotechnol. Bioeng.* 54:503.

Maa, Y., Nguyen, P., Andya, J., Dasovich, N., Sweeney, T., Shire, S., and Hsu, C., 1998a. Effect of spray drying and subsequent processing conditions on residual moisture content and physical/biochemical stability of protein inhalation powders. *Pharm. Res.* 15:768.

Maa, Y., Nguyen, P., and Hsu, C., 1998b. Spray-drying of air-liquid interface sensitive recombinant human growth hormone. *J. Pharm. Sci.* 87:152.

Mansfield, M., 1993. A review of theories of the glass transition, in: *The Glassy State in Foods*, J. Blanschard and P. Lillford, eds., Nottingham University Press.

Masters, K., 1991. *The Spray Drying Handbook*, Longman Scientific, New York.

Mehta, D., Corbo, D., and Iqbal, K., 1996. Spray Dried Erythropoietin. International Patent WO 96/18647.

Millqvist-Fureby, A., Malmsten, M., and Bergenstahl, B., 1999. Spray-drying of trypsin—surface characterization and activity preservation. *Int. J. Pharm.* 188:243.

Mumenthaler, M., Hsu, C., and Pearlman, R., 1994. Feasibility study on spray-drying protein pharmaceuticals: recombinant human growth hormone and tissue-type plasminogen activator. *Pharm. Res.* 11:12.

Niven, R., Lott, F., Ip., A., and Cribbs, J., 1994. Pulmonary delivery of powders and solutions containing recombinant human granulocyte colony-stimulating factor to the rabbit. *Pharm. Res.* 11:1101.

Patton, J., Foster, L., and Platz, R., 1995. Methods and compositions for pulmonary delivery of insulin. International Patent WO 95/24183.

Pikal, M., 1994. Freeze-Drying of Proteins: Process, Formulation and Stability, in: *Formulation and Delivery of Proteins*, J.L. Cleland and R. Langer, eds., American Chemical Society, Washington, DC.

Pikal, M., 1999. Mechanisms of protein stabilization during freeze-drying and storage: The relative importance of thermodynamic stabilization and glassy state relaxation dynamics, in: *Freeze-Drying/Lyophilization of Pharmaceutical and Biological Products*, L. Rey and L.C. May, eds., Marcel Dekker, New York.

Tripp, B., Magda, J., and Andrade, J., 1995. Adsorption of globular proteins at the air/water interface as measured via dynamic surface tension: concentration dependence, mass-transfer considerations and adsorption kinetics. *J. Coll. Interface Sci.* 173:16.

Tzannis, S., and Prestrelski, S., 1999a. Activity-stability considerations of trypsinogen during spray-drying: effects of sucrose. *J. Pharm. Sci.* 88:351.

Tzannis, S., and Prestrelski, S., Moisture effects on protein-excipient interactions in spray-dried powders. Nature of destabilizing effects of sucrose. *J. Pharm. Sci.* 88:360.

7

Surfactant-Protein Interactions

Theodore W. Randolph[1,2] *and LaToya S. Jones*[1,3]

INTRODUCTION

To retain biological activity, proteins generally must be maintained in a specific, three-dimensional conformation. This conformation is only marginally stable, and thus relatively minor perturbing forces can disrupt protein structure, causing loss of biological activity, as well as formation of non-native protein aggregates. Such perturbations are commonly encountered as proteins are produced, stored, transported, and delivered to patients. For example, it is well known that during common industrial processes such as filtering (Maa and Hsu, 1998), storage (Mcleod et al., 2000), agitation (Thurow and Geisen, 1984; Maa and Hsu, 1997) freeze/thawing (Eckhardt, Oeswein et al., 1991; Nema and Avis, 1993; Izutsu et al., 1994), lyophilization (Carpenter and Chang, 1996; Carpenter et al., 1997), nebulization (Ip et al., 1995) and spray-drying (Broadhead et al., 1994; Mumenthaler et al., 1994; Maa et al., 1998; Adler and Lee, 1999; Millqvist-Fureby, Malmsten et al., 1999; Tzannis and Prestrelski, 1999) proteins can suffer damage to their native conformation. Further, delivery of protein pharmaceuticals to patients may also provoke losses of conformational integrity via unfavorable interactions of proteins with surfaces (e.g., inner walls of catheter tubing or syringes (Tzannis et al., 1996)).

Theodore W. Randolph and LaToya S. Jones • Center for Pharmaceutical Biotechnology. *Theodore W. Randolph* • Department of Chemical Engineering, University of Colorado, Boulder, CO 80503. *LaToya S. Jones* • Department of Pharmaceutical Sciences, School of Pharmacy, University of Colorado Health Sciences Center, Denver, CO 80262.

Rational Design of Stable Protein Formulations, edited by Carpenter and Manning. Kluwer Academic / Plenum Publishers, New York, 2002.

The mechanisms of degradation of protein structure and activity are often categorized in two broad classes, chemical and physical. Chemical degradation refers to those modifications involving covalent bonds, such as deamidation, oxidation and disulfide bond shuffling. Physical degradation includes unfolding of the protein, undesired adsorption of the protein to surfaces, and aggregation. The two categories are not completely independent of one another. For example, protein oxidation may result in a greater proclivity to aggregate, and the rate of non-native disulfide bond formation may be higher in aggregated proteins.

Surface-active agents, or surfactants, are often added to protein solutions to prevent physical damage during purification, filtration, transportation, freeze-drying, spray-drying and storage. Surfactants are amphiphilic, containing a polar head group and a non-polar tail. This dual nature causes surfactants to adapt specific orientations at interfaces and in aqueous solutions. It is this characteristic that lies at the root of the mechanisms by which surfactants affect the physical stability of proteins.

A well-known example is the anionic surfactant sodium dodecyl sulfate, or SDS. The sulfate anion is the hydrophilic head group of SDS, while the long aliphatic dodecyl chain forms the tail group. Ionic surfactants such as SDS have been known since the late 1930's as effective protein denaturants (Anson, 1939), and are commonly used for this purpose, e.g., as a pre-treatment for proteins in polyacrylamide gel electrophoresis (SDS-PAGE). In contrast, surfactants used as stabilizing agents in protein formulations are typically non-ionic (Loughheed et al., 1983; Twardowski et al., 1983; Chawla et al., 1985). This chapter will focus on non-ionic surfactants; protein interactions with ionic surfactants have been reviewed elsewhere (Jones, 1996). An example non-ionic surfactant is polyoxyethylene sorbitan monolaurate (Tween 20®), shown in Figure 1. In this molecule, the hydrophilic polyoxyethylene units form the head group, while the hydrophobic monolaurate group is the tail. Tween 20 is often added to formulations due to its ability to protect proteins from surface-induced denaturation (Chang et al., 1996; Jones et al., 1997; Bam et al., 1998; Kreilgaard et al., 1998; Maa et al., 1998).

There are a number of mechanisms by which surfactants can prevent or promote damage to proteins. Some of these mechanisms are generic to all excipients, and can be explained in the solution thermodynamic framework of the Wyman linkage theory (Wyman and Gill, 1990) and the preferential exclusion mechanisms developed by Timasheff and colleagues (Arakawa and Timasheff, 1982, 1983, 1984a,b, 1985a,b,c; Arakawa et al., 1990; Timasheff, 1998). Others derive from the amphiphilicity of surfactants and the resulting effect of microscopic ordering of surfactant molecules at interfaces, which in turn affects the kinetics and thermodynamics of protein interfaces. In this chapter, we discuss a number of these mechanisms and their implications for protein stability.

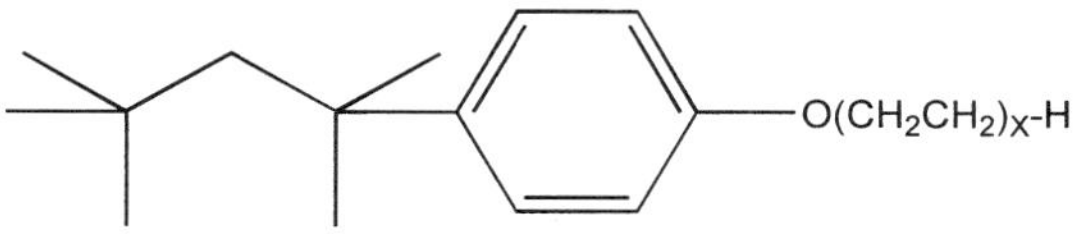

Polyethylene glycol ether
Triton X-100, x=9-10 (average)
Triton X-114, x=7-8 (average)

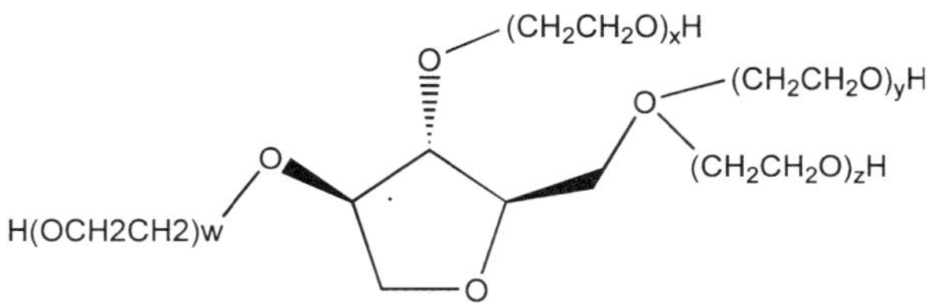

w+x+y+z=20
Polysorbate
Tween 20, $R=C_{11}H_{23}CO_2$
Tween 80, $R=C_{17}H_{33}CO_2$

Figure 1. Example non-ionic surfactants.

PROTEINS AND SURFACTANTS AT SURFACES

Because of their dual hydrophobic/hydrophilic nature, surfactants in solution tend to orient themselves so that the exposure of the hydrophobic portion of the surfactant to the aqueous solution is minimized. Thus, in systems containing air/water interfaces, surfactants will tend to accumulate at these interfaces, forming a surface layer of surfactant oriented in such a fashion that only their hydrophilic ends are exposed to water. Such orientation and surface adsorption can also occur at solid/water interfaces such as those found in vials, syringes, and other containers. Protein molecules also exhibit surface activity (for a review see (Magdassi, 1996), and references therein) and as such will also tend to adsorb to and orient at these interfaces.

From classical thermodynamics, the excess surface internal energy dU_1^σ of a surface with area A at a temperature T is related to the excess surface entropy

S_1^σ and the chemical potential and number of surface excess moles of each adsorbed species:

$$dU_1^\sigma = TdS_1^\sigma + \sigma dA + \sum_{i=2}^{c} \mu_i dn_{i,1}^\sigma \tag{1}$$

Here the subscript 1 refers to a dividing surface chosen so that there is no excess adsorption of species 1, the solvent (water) (Gibbs, 1961), and σ is the surface tension. The equilibrium criterion, $\delta S = 0$, requires that σ be constant across the surface. Thus, if a protein adsorbs to an interface, at equilibrium the surface tension forces must be continuous and constant across the whole interface, including across the protein. The stability criterion at equilibrium requires that:

$$\left(\frac{\partial \sigma}{\partial A}\right)_{T,n} > 0 \tag{2}$$

If the surface tension of the interface is greater than the internal tension in the protein, then in order to meet these two conditions, the surface area of the protein must increase until the two tensions are equal, i.e., the protein must unfold. In some cases, nearly complete loss of native activity is lost upon adsorbing to the interface (Rothen, 1947; Verger et al., 1973). The Gibbs adsorption equation relates the surface tension to the concentration of adsorbed species at an interface:

$$-d\sigma = s_1^\sigma dT + \sum_{i=2}^{c} \Gamma_{i,1} d\mu_i \tag{3}$$

where $\Gamma_{i,1}$ and $s_1^\sigma dT$ are, respectively, the excess surface adsorption and excess surface entropy of component i, both relative to a dividing surface with no surface excess of solvent (1), and μ_i is the chemical potential of species i. Adsorption of protein to the interface thus lowers the interfacial tension, making unfolding less likely as adsorption progresses.

If the process of surface adsorption and unfolding of protein were to stop after the formation of an equilibrium monolayer, the amount of adsorbed protein would be so small as to be generally of no consequence. Indeed, a significant amount of work has been dedicated to development of methods with sensitivities high enough to characterize the minute amount of protein adsorbed to the interface (Tupy et al., 1998; Vermeer and Norde, 2000). However, depending on the degree of surface hydrophobicity and characteristics of the protein in question, additional processes can occur in the adsorbed films, leading to behavior that is no longer described by the Gibbs' adsorption equation. Among the possible processes are gas-to-liquid surface phase transitions, surface precipitation, and the formation of surface sublayers. In many cases, the adsorbed molecules can be rapidly released from the surface and surface sublayers, and may exchange with bulk protein molecules. Alternatively, because of slow refolding kinetics, proteins can become irreversibly adsorbed at the interface, or have a rate of exchange somewhere between these extremes (Dickinson, 1999; Norde and Giacomelli, 1999). These processes of adsorption and release of structurally-perturbed protein molecules into the bulk solution have been implicated as one of the causes of protein aggregation and denaturation.

When discussing protein adsorption at interfaces, globular proteins are typically characterized as being either "hard" or "soft," having a low or high degree of flexibility, respectively, and by their degrees of hydrophobicity. Soft, hydrophobic proteins, attaining monolayer coverage of the air/water interface in the matter of minutes, are generally more surface reactive at hydrophobic surfaces than hard, hydrophilic proteins, attaining coverage of the same surfaces in a matter of hours (Tripp et al., 1995). The driving force for protein adsorption is the decrease in the entropy of the water molecules that are ordered around the hydrophobic protein domains when the protein is in the bulk solution. Thus, the role of the relative degree of hydrophobicity on protein surface adsorption is rather straightforward: given two proteins only differing in their hydrophobicities, the more hydrophobic protein will have a greater number of productive interactions with the surface and will form the monolayer more quickly. Middelberg et al. (2000) proposed that the difference in adsorption kinetics of Lac21 and Lac28 peptides is because the monomeric Lac21 has more hydrophobic residues exposed than its tetratmeric counterpart, Lac28; thus, Lac21 more readily forms a monolayer at an octane-water interface. Protein flexibility is important in protein spreading that occurs at the interface. A flexible protein can expose additional non-polar residues, leading to an increased strength in binding to the surface. Finally, the protein flexibility dictates the number of proteins that can adsorb at the interface, and their spreading rate (Norde and Giacomelli, 1999).

Protein adsorption to a hydrophobic surface does not necessarily lead to a *complete* loss of "native" structure: some proteins actually gain structure. For example, melittin, a honeybee venom peptide, increases its α-helical content slightly when adsorbed to hydrophobic quartz. In contrast, adsorption of a tetramer of the same peptide is thought to require a loosening of the α-helical content. The orientation of the adsorbed helices in the peptide is parallel to the quartz plane, with the hydrophobic moieties facing the plane (Smith and Clark, 1992). Caessens et al. (1999) also report an increase in helical content on the adsorption of the predominately random coil β-casein and β-casein peptides at the teflon/water interface.

Surfactants will also adsorb to the interface. The Gibbs adsorption isotherm again predicts that this interaction will lower the surface tension. Thus, if surfactants co-adsorb to an interface together with proteins, there will be a smaller driving force for surface adsorption. Furthermore, competition between the protein and the surfactant for the interface may reduce equilibrium protein adsorption. For example, Tween 20 addition displaces beta-lactoglobulin films from air-water interfaces (Roth et al., 2000). This effect is not universal: the ionic surfactant sodium dodecyl sulfate forms a complex with high surface activity that exhibits enhanced surface adsorption (Green et al., 2000). Finally, because the surface tension is lowered after surfactant is adsorbed, less damage to the protein that does adsorbed may occur. For example, a loss of α-helix and an increase in β-turn structures occur when bovine serum albumin adsorbs to polystyrene particles. This effect is decreased when the surface is more crowded (Norde and Giacomelli, 2000).

Protein stabilization by nonionic surfactants can often be observed by formulating with micromolar concentrations of surfactant. This is due to the high surface-activity of this class of excipients, which renders a higher effective concentration of surfactant molecules at interfaces than in the bulk solution. When the concentration of the surfactant is much lower than its critical micelle concentration (CMC), surfactant molecules lie flat at the air/water and hydrophobic solid/water interfaces (Figure 2A) (Porter, 1994). As the concentration is increased, more molecules adsorb to these interfaces, such that the surface concentration remains linearly proportional to the bulk concentration (Tanford, 1973). This crowding forces the surfactant molecules to order themselves such that the hydrophilic groups are oriented towards the bulk water and the hydrocarbon chains are pointed towards the air or hydrophobic solid (Porter, 1994; Fainerman et al., 2000) (Figure 2B). At sufficiently high surfactant concentrations (i.e., at or above the CMC), there is an oriented monolayer of surfactant molecules and maximum surfactant absorption, at the interface (Figures 2C & D). The surface saturation is responsible for the sharp slope change (to essentially zero) observed in experimental plots of surface properties (surface tension, osmotic pressure) versus surfactant concentration (Porter, 1994) (Figure 3). It should be noted that the linear variations in the surface properties shown in Figure 3 are indicative that the surface activity of a surfactant is due to the hydrophobic effect: the variation would be cooperative if hydrocarbon self-affinity was the appropriate explanation (Tanford, 1973). Surfactant micelles are formed in the bulk phase when the concentration is of the surfactant is above the CMC (Figure 2D). Thus, a range of surfactant concentrations could inhibit protein denaturation at an interface; however, the necessity of CMC levels of surfactant to completely inhibit protein damage would be a strong indication that the damage is caused by adsorption of protein at the interface and that inhibiting protein adsorption at the interface plays a role in preventing protein aggregation.

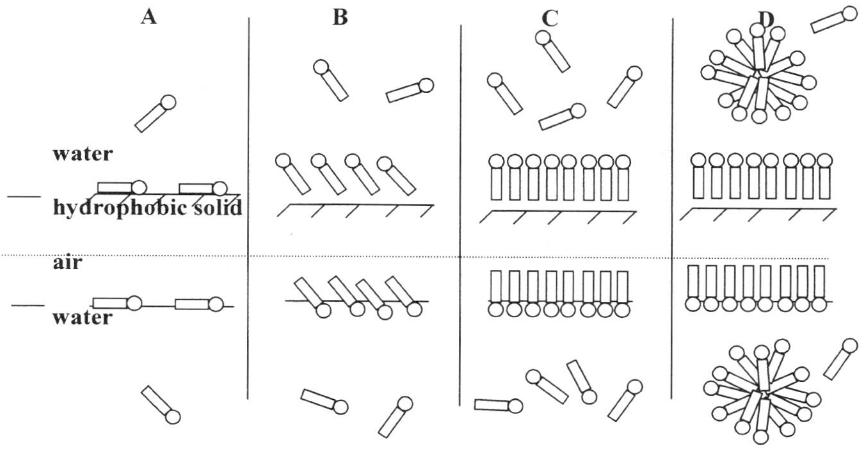

Figure 2. Simplified models of the interfacial behavior of a nonionic surfactant at several concentrations in water. Circles—polar head groups (e.g., polyoxyethelenes). Rectangles—hydrophobic tails (e.g., hydrocarbon chains). Models above the dotted line is for the hydrophobic solid (ice)/water interface and those above are for the air/water interface. (**A**) Surfactant concentration is well below the CMC (**B**) Surfactant concentration is greater than in **A**, but still below the CMC. (**C**) Surfactant concentration is at the CMC. (**D**) The surfactant concentration is above the CMC. (This model is adapted from Figures 4.4 and 4.8 of Porter (Porter, 1994)).

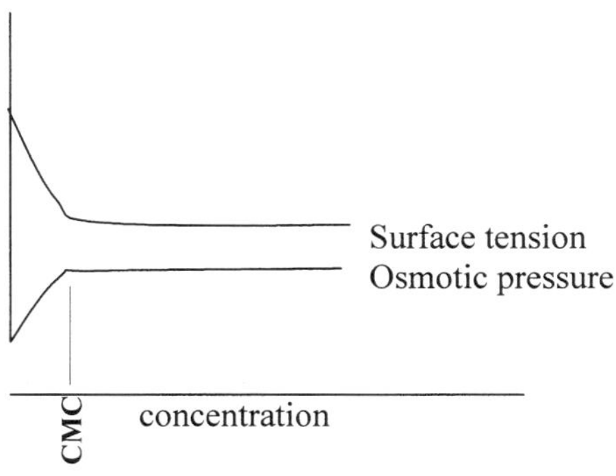

Figure 3. Representation of changes two properties in determining the CMC of a surfactant.

Decreased protein adsorption at interfaces (e.g., air/liquid, ice/liquid) in the presence of nonionic surfactants such as Tween 20 can be attributed to the surface activity of nonionic surfactants (Chang et al., 1996; Kreilgaard et al., 1998; Miller et al., 2000a,b) and, in some cases, direct interactions between the surfactant and protein molecules (Dickinson, 1998; Bam et al., 1998; Miller et al., 2000a,b). In mixed protein/surfactant systems in which the surfactant binds to hydrophobic regions of the protein, the protein is less surface-active than it would be in a solution devoid of the nonionic surfactant. This explains the increase in surface tension relative to that of the pure protein solution that can be observed at extremely low surfactant concentrations. Adsorption of surfactant and protein molecules in the mixed system is competitive. Nonionic surfactants usually bind tighter than proteins or protein–surfactant complexes at interfaces (Dickinson, 1998). Thus, above a critical concentration of the surfactant, protein adsorption becomes negligible and the adsorption isotherms for mixed surfactant/protein systems can be roughly identical that of a pure surfactant solution, as observed by Miller et al. (2000a) for the HSA/C_{10}DMPO when the C_{10}DMPO concentration exceeds 10^{-7} mol/cm^3.

PROTEIN-SURFACTANT INTERACTIONS IN SOLUTION

In addition to altering the interaction of proteins with surfaces, non-ionic surfactants can also interact directly with proteins in solution. For example, Tween 20 acts as a chemical chaperone, aiding in the refolding of proteins (Bam et al., 1998; Kreilgaard et al., 1999). *In vivo*, proteins fold while at average concentrations of approximately 35 mg/ml (Hartl, 1996): *in vitro*, non-native protein molecules at this concentration (e.g., due to freeze concentration) usually aggregate. Inside cells, protein folding is aided by naturally occurring molecular chaperones. Unlike folding catalysts that have steric information to guide the protein folding, molecular chaperones act by non-covalently binding to partially-folded proteins, usually via hydrophobic interactions, to prevent misfolding or aggregation while the protein is attempting to adopt its native conformation (Gatenby and Ellis, 1990; Hartl, 1996). Chaperone-assisted protein refolding can prevent the protein from falling into kinetic traps or simply allow more time for the protein to refold (Gatenby and Ellis, 1990). The hydrophobic effect, the driving force of protein folding and surface activity of Tween 20, is also implicated in the interactions between the exposed hydrophobic regions of the partially folded protein and the hydrocarbon tail of the surfactant. These mixed surfactant/protein complexes and protein folding are dynamic processes, eventually the protein attains a conformation in which the hydrophobic groups are not as surface exposed. Unless there are hydrophobic patches in the native protein conformation, the surfactant molecules will not necessarily bind to the protein, as in the case of molecular chaperones. *In vitro*, surfactants such as Tweens (Bam et al., 1996), polyethylene glycol (Cleland and Wang, 1990; Cleland et al., 1992; Cleland and Randolph, 1992; Cleland 1993; Cleland and Wang, 1993), Triton

X-100 (Donate et al., 1998), and lubrol (Donate et al., 1998), have been implicated in aiding protein refolding by acting as chemical chaperones.

Surfactants, such as Tween 20, can also affect the thermodynamic conformational stability of a protein. As discussed in other chapters in this book, and in detail in references from Timasheff and co-workers (Lee and Timasheff, 1981; Arakawa and Timasheff, 1982; Arakawa and Timasheff, 1985; Timasheff, 1998), thermodynamic stability is increased if a ligand exhibits greater binding to the native state of a protein than to a non-native state. However, with many excipients (e.g., sucrose) excluded volume effects produce a non-specific negative binding to the native state, and a concomitantly larger negative binding to expanded, non-native conformations. This differential negative binding also results in a stabilization of the native state. At the low concentrations of surfactant (*ca.* 100 micromolar) typically used in formulations of therapeutic proteins, thermodynamic effects due to excluded volume can usually be neglected. More important is specific binding of surfactants to either the native or unfolded states of a protein. Randolph and colleagues report that some proteins and nonionic surfactants, including Tween 20, form mixed protein: detergent complexes (Bam et al., 1995, 1996, 1998; Jones, et al., 1999). In the presence of Tween 40 at a 4:1 surfactant: protein molar ratio, native recombinant human growth hormone is significantly stabilized: the denaturation midpoint in guanidine hydrochloride solutions increases from 4.6 M guanidine hydrochloride in the absence of Tween 40 to 5.9 M in the presence of Tween 40 (Bam et al., 1996). Likewise, 4:1 Tween 40 increases the ΔG of unfolding of recombinant human growth hormone by 4.1 kcal/mol (Bam et al., 1996), and 10:1 Tween 40 increases its melting point slightly from 88.8 to 89.4° C (Bam et al., 1998). Stabilization of recombinant human growth hormone by surfactants results in reduced aggregation in agitated solutions (Bam et al., 1998). With bovine serum albumin in the presence of surfactant there are decreased amounts of thermally-induced protein aggregates, relative to surfactant-free controls (Arakawa and Kita, 2000). Conformational stabilization of proteins by non-ionic surfactants is not universal; stability of IgG is unaffected by low concentrations of Tween 20 (Vermeer and Norde, 2000), and recombinant human interferon-γ shows lower free energies of unfolding in the presence of Tween 20 (Webb et al., 2000).

SURFACTANT EFFECTS ON PROTEIN ASSEMBLY STATE

The hydrophobic portion of non-ionic surfactants can bind to hydrophobic patches on proteins. This naturally causes the surfactant to order itself so that more hydrophilic groups are solvent exposed, resulting in a "hydrophobicity reversal". This "hydrophobicity reversal" means that the protein-surfactant complex is more hydrophilic that either the surfactant or protein alone, and effectively increases the solubility of the complex. This, in turn can reduce the propensity of the protein to form higher-order aggregates. For example, bovine mitochondrial cytochrome bc1 is dimeric in solutions at low ionic strength and low surfactant levels. At Tween 20 concentrations above 5 mg/mg protein, a homogeneous, monomeric, reversible and enzymatically active protein-surfactant complex is formed (Musatov and Robinson, 1994). Likewise, in freeze-thaw studies of recombinant factor XIII, addition of Tween 20 at concentrations near the CMC blocked the progression of aggregates from a relatively low molecular weight, soluble fraction to insoluble aggregates. Figure 4 shows levels of

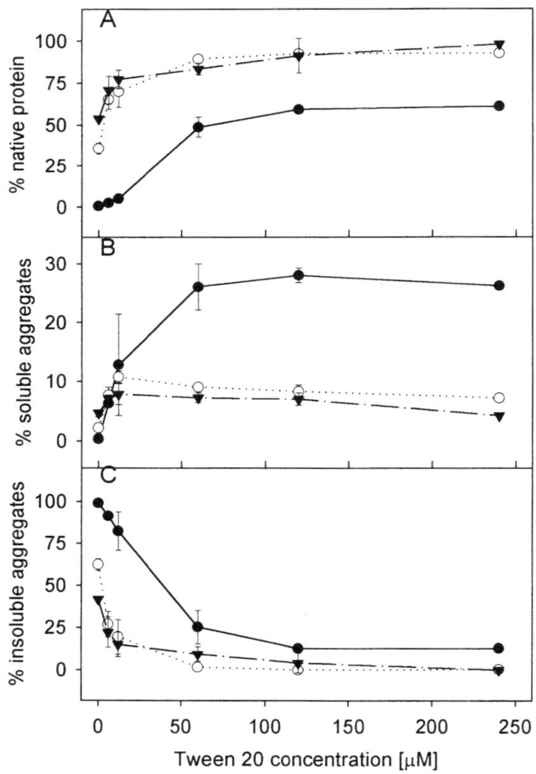

Figure 4. Recovery of native rFXIII (A) and formation of soluble (B) and insoluble aggregates (C) following 10 freeze-thaw cycles of 1 mg/ml (—●—), 5 mg/ml (·····○·····) and 10 mg/ml rFXIII (—▼—) as a function of Tween 20. Results are plotted as mean values +/− standard deviation for duplicate samples (reproduced from Kreilgaard et al., 1998).

aggregation for factor XIII after freeze thaw cycling in the presence and absence of Tween 20. Note that Tween addition did not completely block aggregation, but was very effective at preventing the formation of insoluble aggregates (Kreilgaard et al., 1998).

Non-ionic surfactants can also have the opposite effect on protein assembly state. In cases where a non-ionic surfactant destabilizes the conformation of a protein, this effect may compete against the solubilizing effect of surfactant binding and hydrophobicity reversal. For example, Bax, a monomeric protein that regulates apoptosis, readily forms dimers in the presence of Tween 20. However, these dimers are apparently non-native, as they do not expose the characteristic N-terminal Bax epitope (Hsu and Youle, 1998). In the case of the hydrophobic lipase from *Humicola lanuginose*, Tween 20 addition caused the formation of large, insoluble non-native aggregates (Kreilgaard et al., 1999).

SURFACTANT EFFECTS ON PROTEINS DURING FREEZING, FREEZE-DRYING AND RECONSTITUTION

The processes of freezing, drying, and reconstitution of protein solutions present a number of stresses that may denature proteins. Many of these stresses are associated with surfaces: new ice-water and ice-glassy solid interfaces are formed during freezing, drying replaces ice-glass interfaces with air-glass interfaces, and reconstitution exposes the glassy solid surfaces to aqueous solution. In each of these steps protein adsorption to surfaces is potentially damaging. The ice-water interface has been implicated as a source of damage to proteins (Strambini and Gabellieri, 1996), as has the solid-air interface (Hsu et al., 1995). Addition of nonionic surfactants can reduce this damage, presumably by competing with the protein for the ice-water interface (Chang, 1996). For example, addition of Tween 80 to solutions of recombinant hemoglobin reduced aggregation seen during freeze thaw studies (Kerwin et al., 1998). Interestingly, Tween 80 did not offer protection against methemoglobin formation or hemoglobin aggregation during long-term frozen storage.

Non-ionic surfactants also have been shown to affect the recovery of native protein from lyophilized formulations. Sarciaux et al. (1999) showed that the addition of Tween 80 to the formulation solution or the reconstitution medium for lyophilized formulations resulted in reduced levels of aggregates. Likewise, Zhang et al. (1995, 1996) have demonstrated that, following long-term storage, surfactants in the reconstitution medium can affect protein recovery. We have recently shown (Webb et al., 2000) that addition of Tween 20 to the reconstitution medium for lyophilized preparations of recombinant human interferon-γ results in decreased levels of aggregates. The mechanism for such reduced aggregation was shown to be a surfactant effect on dissolution rates. Addition of Tween 20 slowed the dissolution of the lyophilized solid, allowing protein that partially unfolded during freeze-drying to refold before aggregating. Interestingly, Tween 20 in aqueous solutions destabilizes recombinant human interferon-γ against urea-induced unfolding, impedes refolding during rapid dilution from urea solutions, and actually increases aggregation during agitation.

ENZYMATIC DEGRADATION OF NON-IONIC SURFACTANTS

Although most non-ionic surfactants are thought of as chemically inert components of a formulation, specific chemical interactions between proteins can occur. For example, some enzymes show hydrolytic activity toward Tweens. Smegmatocin, an esterase from *Mycobacterium smegmatis*, shows a broad thermal and pH stability in its activity against Tween 80 (Tomioka, 1983). The byproduct of Tween 80 hydrolysis, oleic acid, is toxic to some bacteria. Similar bacteriocins, which require Tween their expression, have been ascribed to other mycobacteria (Saito et al., 1983). It is not clear how widespread esterase activity is against Tweens. However, it is clear that caution should be used when formulating proteins with esterase activities in Tween solutions.

RECOMMENDATIONS FOR PROTEIN FORMULATION

Clearly, it is desirable to minimize the addition of any excipient to a formulation. This rule of thumb is even more pertinent for surfactants, because there is ample evidence that high concentrations of surfactants can be destabilizing to protein structure. On the other hand, small amounts of surfactant often provide benefits in preventing aggregation that greatly outweigh any conformationally destabilizing effect. How then should surfactant levels be chosen for optimal formulation? The answer appears to depend on the mechanism(s) by which a particular protein is protected from damage by surfactant addition. In cases where surfactants act to stabilize the native state of a protein by binding to the protein, a specific surfactant : protein stoichiometry may need to be maintained in order to provide optimal protection. In these cases, changes in the protein concentration within a formulation will dictate proportional changes in surfactant concentration to maintain a fixed molar ratio. This appears to be the case for recombinant human growth hormone, where protection against agitation-induced damage correlated with the molar ratio of surfactant to protein rather than to the surfactant's CMC (Bam et al., 1998). In the case of specific binding, the choice of nonionic surfactant may be important. In the case of recombinant human growth hormone, for example, different binding stoichiometries and degrees of protein stabilization were seen for a variety of common surfactants (Bam et al., 1995). A general recommendation for proteins that show specific binding to the native state of the protein is to formulate so that the ratio of surfactant to protein is slightly above the binding stoichiometry for a particular surfactant. The choice of surfactant may be dictated by the degree of stabilization (which should correlate with the degree of binding) provided to a protein by a particular surfactant.

In contrast, if no specific binding is seen, then maximum levels of protection generally correlate with the CMC of the surfactant. In this case, surfactant should be added at levels slightly above the CMC. The choice of surfactant is often dictated by a trade-off: surfactants with lower CMC's will require less surfactant in solution to saturate surfaces and reduce surface-induced damage to proteins. However, surfactants with low CMC's are much more difficult to remove from solution (e.g., by dialysis) if necessary, and also tend to be less soluble than surfactants with higher CMC's, raising the possibility of undesirable phase separation during processes such as freezing or lyophilization.

REFERENCES

Adler, M., and Lee, G., 1999. Stability and surface activity of lactate dehydrogenase in spray-dried trehalose. *Journal of Pharmaceutical Sciences* 88:199.

Anson, M., 1939. The denaturation of proteins by detergents and bile salts. *Science* 90:256.

Arakawa, T., Bhat, R. et al., 1990. Preferential interactions determine protein solubility in three-component solutions: the $MgCl_2$ system. *Biochemistry* 29:1914.

Arakawa, T., Bhat, R. et al., 1990. Why preferential hydration does not always stabilize the native structure of globular proteins. *Biochemistry* 29:1924.

Arakawa, T., and Kita, Y., 2000. Protection of bovine serum albumin from aggregation by Tween 80. *J. Pharm. Sci.* 89:646.

Arakawa, T., and Timasheff. S.N., 1982. Preferential interactions of proteins with salts in concentrated solutions. *Biochemistry* 21:6545.

Arakawa, T., and Timasheff, S.N., 1982. Stabilization of protein structure by sugars. *Biochemistry* 21:6536.

Arakawa, T., and Timasheff, S.N., 1983. Preferential interactions of proteins with solvent components in aqueous amino acid solutions. *Arch. Biochem. Biophys.* 224:169.

Arakawa, T., and Timasheff, S.N., 1984a. Mechanism of protein salting in and salting out by divalent cation salts: balance between hydration and salt binding. *Biochemistry* 23:5912.

Arakawa, T., and Timasheff, S.N., 1984b. Protein stabilization and destabilization by guanidinium salts. *Biochemistry* 23:5924.

Arakawa, T., and Timasheff, S.N., 1985a. Mechanism of poly(ethylene glycol) interaction with proteins. *Biochemistry* 24:6756.

Arakawa, T., and Timasheff, S.N., 1985b. The stabilization of proteins by osmolytes. *Biophys. J.* 47:411.

Arakawa, T., and Timasheff, S.N., 1985c. Theory of protein solubility. *Meth. Enzymol.* 114:49.

Bam, N.B., Randolph, T.W., and Cleland, J.L., 1995. Stability of protein formulations: Investigation of surfactant effects by a novel EPR spectroscopic technique. *Pharm. Res.* 12:2.

Bam, N.B., Cleland, J.L. et al., 1996. Molten globule intermediate of recombinant human growth hormone: stabilization with surfactants. *Biotech. Prog.* 12:801.

Bam, N.B., Cleland, J.L. et al., 1998. Tween protects recombinant human growth hormone against agitation-induced damage via hydrophobic interactions. *J. Pharm. Sci.* 87:1554.

Bam, N.B., Randolph, T.W. et al., 1995. Stability of protein formulations: investigation of surfactant effects by a novel EPR spectroscopic technique. *Pharm. Res.* 12:2.

Broadhead, J., Rouan, S.K. et al., 1994. The effect of process and formulation variables on the properties of spray-dried beta-galactosidase. *J. Pharmaceut. Pharmacol.* 46:458.

Caessens, P.W., De Jongh, H. et al., 1999. The adsorption-induced secondary structure of β-casein and of distinct parts of its sequence in relation to foam and emulsion properties. *Biochim. Biophys. Acta* 1430:73.

Carpenter, J.F., and Chang, B.S., 1996. Lyophilization of protein pharmaceuticals, in: *Biotechnology and Biopharmaceutical Manufacturing, Processing, and Preservation*, K.E. Avis and V.L. Wu, eds., Interpharm Press, Buffalo Grove, IL.

Carpenter, J.F., Pikal, M.J., Chang, B.S., and Randolph, T.W., 1997. Rational design of stable lyophilized protein formulations: some practical advice. *Pharm. Res.* 14:969.

Chang, B.S., Kendrick, B.S., and Carpenter, J.F., 1996. Surface-induced denaturation of proteins during freezing and its inhibition by surfactants. *J. Pharm. Sci.* 85:1325.

Chawla, A.S., Hinsberg, I. et al., 1985. Aggregation of insulin, containing surfactants in contact with different materials. *Diabetes* 34:420.

Cleland, J. L., 1993. Impact of protein folding on biotechnology. in: *Protein Folding. In Vivo and In Vitro.* J.L. Cleland, ed., American Chemical Society, Washington, D.C.

Cleland, J.L., Hedgepeth, C. et al., 1992. Polyethylene glycol enhanced refolding of bovine carbonic anhydrase B. *J. Biol. Chem.* 267:13327.

Cleland, J.L., and Randolph, T.W., 1992. Mechanism of polyethylene glycol interaction with the molten globule folding intermediate of bovine carbonic anhydrase B. *J. Biol. Chem.* 267:3147.

Cleland, J.L., and Wang, D.I.C., 1990. Cosolvent assisted protein refolding. *Bio/Technology* 8:1274.

Cleland, J.L., and Wang, D.I.C., 1993. Cosolvent effects on refolding and aggregation. in: *Biocatalyst Design for Stability and Specificity*, M.E. Himmel and G. Georgiou, eds., American Chemical Society, Washington, DC.

Dickinson, E., 1998. Proteins at interfaces and in emulsions. *J. Chem. Soc., Faraday Trans.* 94:1657.

Dickinson, E., 1999. Adsorbed protein layers at fluid interfaces: interactions, structure and surface rheology." *Colloids and Surfaces B: Biointerfaces* 15:161.

Donate, F., Artigues, A. et al., 1998. Opposite behavior of two isozymes when refolding in the presence of non-ionic detergents. *Protein Science* 7:1811.

Eckhardt, B.M., Oeswein, J.Q. et al., 1991. Effect of freezing on aggregation of human growth hormone. *Pharm. Res.* 8:1360.

Fainerman, V.B., Miller, R. et al., 2000. Effect of surfactant interfacial orientation/aggregation on adsorption dynamics. *Adv. Coll. Inter. Sci.* 86:83.

Gibbs, J., 1961. *The Scientific Papers of J. Willard Gibbs.*, Dover Publications, New York.

Green, R., Su, T. et al., 2000. Interaction of lysozyme and sodium dodecyl sulfate at the air-liquid interface. *Langmuir* 16:5797.

Hartl, F.U., 1996. Molecular chaperones in cellular protein folding. *Nature* 381:571.

Hsu, C.C., Nguyen, H.M. et al., 1995. Surface denaturation at solid-void interface—a possible pathway by which opalescent particulates form during the storage of lyophilized tissue-type plasminogen activator at high temperatures. *Pharm. Res.* 12:69.

Hsu, Y., and Youle, J., 1998. Bax in murine thymus is a soluble monomeric protein that displays differential detergent-induced conformations. *J. Biol. Chem.* 273:10777.

Ip, A., Arakawa, T. et al., 1995. Stability of recombinant consensus interferon to air-jet and ultrasonic nebulization. *J. Pharm. Sci.* 84:1210.

Izutsu, K., Yoshioka, S. et al., 1994. Stabilizing effect of amphiphilic excipients on the freeze-thawing and freeze-drying of lactate dehydrogenase. *Biotech. Bioeng.* 43: 1102.

Jones, L.S., Bam, N.B. et al., 1997. Surfactant-stabilized protein formulations: A review of protein-surfactant interactions and novel analytical methodologies. in: *Therapeutic Protein and Peptide Formulation and Delivery.* Z. Shahrokh, V. Sluzky, J. L. Cleland, S. J. Shire and T. W. Randolph, eds. American Chemical Society, Washington, D.C.

Jones, L.S., Cipolla, D. et al., 1999. Investigation of protein-surfactant interactions by analytical ultracentrifugation and electron paramagnetic resonance: the use of recombinant human tissue factor as an example. *Pharm. Res.* 16:808.

Jones, M., 1996. Protein-Surfactant Interactions. in: *Surface Activity of Proteins.* S. Magdassi, ed., Marcel Dekker, New York.

Kerwin, B.A., Heller, M.C. et al., 1998. Effects of Tween 80 and sucrose on acute short-term stability and long-term storage at –20 degrees C of a recombinant hemoglobin. *J. Pharm. Sci.* 87:1062.

Kreilgaard, L., Frokjaer, S. et al., 1999. Effects of additives on the stability of Humicola lanuginosa lipase during freeze-drying and storage in the dried solid. *J. Pharm. Sci.* 88:281.

Kreilgaard, L., Jones, L.S. et al., 1998. Effect of Tween 20 on freeze-thawing- and agitation-induced aggregation of recombinant human factor XIII. *J. of Pharm. Sci.* 87:1597.

Lee, J.C., and Timasheff, S.N., 1981. The stabilization of proteins by sucrose. *J. Biol. Chem.* 256:7193.

Loughheed, W.D., Albisser, A.M. et al., 1983. Physical stability of insulin formulations. *Diabetes* 34:424.

Maa, Y.F., and Hsu, C.C., 1997. Protein denaturation by combined effect of shear and air-liquid interface. *Biotech. Bioeng.* 54:503.

Maa, Y.F., and Hsu, C.C., 1998. Investigation of fouling mechanisms for recombinant human growth hormone sterile filtration. *J. Pharm. Sci.* 87:808.

Maa, Y.F., Nguyen, P.A., 1998. Spray-drying of air-liquid interface sensitive recombinant human growth hormone. *J. Pharm. Sci.* 87:152.

Magdassi, S., 1996. *Surface Activity of Proteins- Chemical and Physicochemical Modifications.*, Marcel Dekker, New York.

Mcleod, A., Walker, I., 2000. Loss of Factor VIII activity during storage in PVC containers due to adsorption. *Haemophilia* 6:89.

Middelberg, A.P.J., and Radke, C.J., 2000. Peptide interfacial adsorption is kineically limited by the thermodynamic stability of self-association. *Proc. Nat. Acad. Sci. USA* 97:5054.

Miller, R., Fainerman, V.B. et al., 2000a. Dynamics of protein and mixed protein/surfactant adsorption layers at the water/fluid interface. *Adv. Coll. Inter. Sci.* 86:39.

Miller, R., Fainerman, V.B. et al., 2000a. Adsorption characteristics of mixed monolayers of a globular protein and a non-ionic surfactant. *Colloids and Surfaces A-Physicochemical and Engineering Aspects* 161:151.

Millqvist-Fureby, A., Malmsten, M. et al., 1999. Spray-drying of trypsin—surface characterisation and activity preservation. *Int. J. Pharm.* 188:243.

Mumenthaler, M., Hsu, C.C. et al., 1994. Feasibility study on spray-drying protein pharmaceuticals: recombinant human growth hormone and tissue-type plasminogen activator. *Pharm. Res.* 11:12.

Musatov, A., and Robinson, N., 1994. Detergent-solubilized monomeric and dimeric cytochrome bc1 isolated from bovine heart. *Biochemistry* 33:13005.

Nema, S., and Avis, K.E., 1993. Freeze-thaw studies of a model protein, lactate dehydrogenase, in the presence of cryoprotectants. *J. Parent. Sci. Tech.* 47:76.

Norde, W., and Giacomelli, C.E., 1999. Conformational changes in proteins at interfaces: from solution to the interface, and back. *Macromol. Symp.* 145:125.

Norde, W., and Giacomelli, C.E., 2000. BSA structural changes during homomolecular exchange between the adsorbed and dissolved states. *J. Biotechnol.* 79:259.

Porter, M.R., 1994. *Handbook of Surfactants.* Blackie Academic & Professional, Glasgow.

Roth, S., Murray, B., et al., 2000. Interfacial shear rheology of aged and heat treated beta-lactoglobulin films: Displacement by non-ionic surfactant. *J. Agri. Food Chem.* 48:1491.

Rothen, A. (1947). *Adv. Protein Chem.* 3:123.

Saito, H., Tomioka, H. et al., 1983. Mycobacteriocins produced by rapidly growing mycobacteria are Tween-hydrolzing esterases. *J. Bacteriol.* 153:1294.

Sarciaux, J.-M., Mansour, S., 1999. Effects of buffer composition and processing conditions on aggregation of bovine IgG during freeze-drying. *J. Pharm. Sci.* 88: 1354.

Smith, L.J., and Clark, D.C., 1992. Measurement of the secondary structure of adsorbed protein by circular dichroism. 1. Measurements of the helix content of adsorbed melitten. *Biochim. Biophys. Acta* 1121:111.

Strambini, G.B., and Gabellieri, E., 1996. Proteins in frozen solutions: evidence of ice-induced partial unfolding. *Biophys. J.* 70:971.

Tanford, C., 1973. *The Hydrophobic Effect: Formation of Micelles and Biological Membranes*, John Wiley & Sons, New York.

Thurow, H., and Geisen, K., 1984. Stabilization of dissolved proteins against denaturation at hydrophobic interfaces. *Diabetologia* 27:212.

Timasheff, S.N., 1998. Control of protein stability and reactions by weakly interacting cosolvents: the simplicity of the complicated. *Adv. Prot. Chem.* 51:355.

Tomioka, H., 1983. Purification and characterization of the tween-hydrolyzing esterase of *Mycobacterium smegmatis. J. Bacteriol.* 155:1249.

Tripp, B.C., Magda, J.J. et al., 1995. Adsorption of globular proteins at the air/water interface as measured via dynamic surface tension: concentration dependence, mass-transfer considerations, and adsorption kinetics. *J. Coll. Inter. Sci.* 173:16.

Tupy, M., Blanch, H. et al., 1998. Total internal reflection fluorescence spectrometer to study dynamic adsorption phenomena at liquid/liquid interfaces. *Indust. Engin. Chem. Res.* 37:3159.

Twardowski, Z.J., Nolph, K.D. et al., 1983. Nature of insulin binding to plastic bags. *Am. J. Hosp. Pharm.* 40:579.

Tzannis, S.T., Hrushesky, W.J.M. et al., 1996. Irreversible inactivation of interleukin-2 in a pump-based delivery environment. *Proc. Nat. Acad. Sci. USA* 85:586.

Tzannis, S.T., and Prestrelski, S.J., 1999. Activity-stability considerations of trypsinogen during spray drying: Effects of sucrose. *J. Pharm. Sci.* 88:351.

Verger, R., Mieras, M.C.E. et al., 1973. *J. Biol. Chem.* 248:4023.

Vermeer, A.W.P., and Norde, W., 2000. CD Spectroscopy of proteins adsorbed at flat hydrophilic quartz and hydrophobic teflon surfaces. *J. Coll. Inter. Sci.* 225:394.

Vermeer, A.W.P., and Norde, W., 2000. The influence of binding of low molecular weight surfactants on the thermal stability and secondary structure of IgG. *Colloids and Surfaces A-Physicochemical and Engineering Aspects* 161:139.

Webb, S., Cleland, J.L. et al., 2000. Tween 20 inhibits aggregation of lyophilized recombinant human interferon-γ by slowing dissolution during reconstitution. *J. Pharm. Sci.* (submitted).

Wyman, J., Jr., and Gill, S.J., 1990. *Binding and Linkage. Functional Chemistry of Biological Macromolecules.*, University Science Books, Mill Valley, CA.

Zhang, M.Z., Pikal, K. et al., 1996. The effect of the reconstitution medium on aggregation of lyophilized recombinant interleukin-2 and ribonuclease A. *Pharm. Res.* 13:643.

Zhang, M.Z., Wen, J. et al., 1995. A new strategy for enhancing the stability of lyophilized protein: the effect of the reconstitution medium on keratinocyte growth factor. *Pharm. Res.* 12:1447.

8

High Throughput Formulation: Strategies for Rapid Development of Stable Protein Products

Rajiv Nayar and Mark C. Manning

INTRODUCTION

With the growing demand for new and innovative medicines, drug companies are spending record amounts of money on research and development. In the U.S. alone, research and development (R&D) investments are expected to exceed $24 billion in 1999, with investments in biotechnology totaling about $7 billion (Pharmaceutical Research and Manufacturers of America, 1999). As a result, new technologies are increasing the efficiency of the drug discovery process, and the drug pipelines have more products than ever in development. Nearly 350 biotechnology-related products are currently in clinical trials and over 50 are on the market (Pharmaceutical Research and Manufacturers of America, 1999). Pharmaceutical companies can now typically assess potential activities of up to 100,000 compounds a day using high throughput screening systems. In the biotechnology arena, recent advances in genomics, functional genomics, proteomics, bioinformatics and pharmacogenomics are facilitating the development of protein drug candidates at a much faster rate than was possible during the early years of the biotechnology industry. Couple these changes with the impending

---

Rajiv Nayar • HTD Biosystems, Hercules, CA. *Mark C. Manning* • Center for Pharmaceutical Biotechnology, University of Colorado Health, Sciences Center, Denver, CO 80262.

Rational Design of Stable Protein Formulations, edited by Carpenter and Manning. Kluwer Academic / Plenum Publishers, New York, 2002.

publication of the complete sequence of the human genome by 2001 and there is a potential for an additional 15,000 protein drugs from the predicted 150,000 potential genes in the human genome.

This new wave of protein drugs will include compounds that can function as growth factors, act as specific stimulators or suppressors of certain functions, or exhibit activities that have not been observed before. Many experts anticipate a golden age of protein-based therapeutics in the twenty-first century. As various research enterprises gear up to search for these new candidates using high throughput cellular and molecular screening assays, there undoubtedly will be a need for a corresponding ability to formulate these drugs much more quickly than in the past. In addition, there will be increased use of novel delivery systems to administer these protein therapeutics efficiently, making stabilization even more challenging. In short, the pharmaceutical biotechnology industry must find new ways to speed up the drug development process, in order to keep pace with the drug discovery process. We think that the limiting factor for introduction of new biotechnology-derived products will not be finding new drug candidates, but efficient and rapid formulation of these compounds into acceptable dosage forms for delivery and testing in humans.

The potential of many new drug candidates flowing through the pipeline will require implementation of an entirely new approach towards product development. The particular challenge addressed here is the demand that will be placed upon formulation groups, where there will be increased expectations to develop stable formulations of:

more proteins,
in a shorter period of time,
with less material, and
with little, if any, increase in personnel.

Therefore, one must consider new strategies for meeting such challenges.

This chapter is intended to serve as a forum for discussing and evaluating what factors must be considered in trying to implement new approaches to formulating protein pharmaceuticals in a rapid, rational fashion. As a comprehensive strategy for developing stable formulations of protein pharmaceuticals is not currently available, we have assembled what we consider to be the key elements necessary to accomplish this critical task. This chapter should not be viewed as a finished product, but as a work-in-progress, encouraging pharmaceutical scientists to comment on the concept and to modify the approach to their particular situations.

Again, the overarching goal is to possess a unified strategy for accommodating many more proteins at one time, while meeting shortening timelines. It should be expected that there would be a limited supply of well-characterized material and a fixed personnel head count. We envision a scheme that we term "high throughput formulation" or HTF. It is based upon our current understanding regarding stabilization of proteins, but it also identifies areas where technical improvements must be made, especially in the area of analytical methodology and software. If the HTF strategy is successful, it should be possible to match the output of stable formulations by development groups with potential leads by discovery groups.

OVERALL STRUCTURE OF THE HTF APPROACH

The HTF concept includes four key elements, which are briefly outlined in Table 1. The heart of the HTF strategy is a decision plan, where the formulation scientist must decide on (a) the specific excipients to be used as well as (b) the final dosage form. Central to the HTF scheme is the use of a limited set of additives as well as dosage forms. Especially for lyophilized or freeze-dried formulations, our understanding of the behavior is such that some standard formulations can be envisioned (see Chapter 5). Also, the choice of basic solution conditions will be based upon preformulation studies that examine the behavior of the protein as a function of pH, ionic strength, and buffer type (see Chapters 1 and 3).

The second key aspect is an increased role for software and information databases. The preformulation studies could require a significant number of independent experiments to be conducted. Therefore, experimental design software should be used to reduce the number of experimental trials and optimize the results through statistical evaluation. In addition, there are a number of databases and associated programs for assessing physical and chemical properties based upon primary sequence of the protein through a number of websites (e.g. SWISS-PROT). These can provide insight into protein structure and other critical characteristics (e.g., pI, hydrophobicity and hydroflexibility plots and overall charge), and even identify residues that might be sensitive to chemical decomposition. We believe such capabilities will expand over time, and will play an increasingly important role in setting a rational plan for preformulation studies. Finally, we anticipate that specific protein stability databases will emerge, allowing scientists to gather data on specific development issues, without searching all of the open literature. Because information on protein stability comes from a multitude of disciplines (e.g., biochemistry, biophysics, food science, engineering, pharmaceutics, surface science and polymer science), open searches often miss crucial publications and documents. The challenge is not unlike Internet search engines trying to find a very specific subset of information on an ever-increasing world wide web. As these resources become more available and refined, they will emerge as invaluable tools for development of pharmaceutically acceptable formulations.

Table 1.
Key Elements of the HTF Approach

1. Decision Plan to Finalize Choice of Excipients and Dosage Form
   - A. Choice of Dosage Form
     1. Frozen Solution
     2. Liquid Formulation
     3. Lyophilized Formulation
     4. Protein Suspension
     5. Specialized Drug Delivery System
   - B. Choice of Excipients
     1. Proper pH/Buffer Species
     2. Salt
     3. Stabilizer
     4. Bulking Agent (for lyophilized formulations)
     5. Surfactant
     6. Chelators
2. Use of Software and Databases to Assist in Formulation Development
   - A. Experimental Design Software
   - B. Software and Databases for Analysis of Structure/Function/Properties/Stability
   - C. Protein Stability Databases
3. Availability of Essential Analytical Methods
4. Design of Appropriate Stability Testing Protocols

The third aspect of HTF is assembling the correct analytical methods to assay the stability, both chemical and physical, of a wide range of proteins. The mechanisms by which proteins degrade have been widely reviewed (Manning et al., 1989; Pearlman and Nguyen, 1992; Cleland et al., 1993), so they will not be discussed here. However, given this backdrop, we will describe certain key methods that must be available in order to conduct proper HTF studies. All of the required instruments do not necessarily need to be available in-house, as many can be accessed through contract organizations and through academic collaborations. In addition, it should be noted that there exists a tremendous opportunity to modify existing analytical techniques for use with smaller sample sizes and for rapid, parallel sampling. One of the expected limitations will be lack of well-characterized material. Whereas access to a minimum of 200–500 mg of protein for formulation studies is desirable, the genomic approach will tax the ability of process scientists to provide adequate amounts of purified proteins for all compounds under development. Therefore, in some cases, the total amount of protein available may be an order of magnitude less than the optimum. This situation should serve as an incentive to identify new approaches for gathering stability data. We will discuss below possible new methodologies that would be useful in the HTF scheme, but have not yet been applied to formulation development.

Finally, one must have a clear plan for stability testing, along with minimum criteria for a "stable" formulation. Certainly, that definition may change as a product matures. Minimum stability for the initial clinical studies may be quite different than the expectations for the final formulation tested in Phase III. Presently, there are debates within companies on how robust a preliminary formulation development or purification process must be in order to demonstrate efficacy in humans. Some companies will accept any minimal formulation in order to initiate clinical trials. However, even these formulations would likely have to have at least six to twelve months of shelf life. Our goal in the HTF approach is to develop a final formulation as early as possible in the drug development and testing process. If this goal can be accomplished, then fewer resources will need to be diverted to redesign the formulation later.

ROLE OF AN ESTABLISHED DECISION TREE FOR FORMULATION DESIGN

The central feature in the HTF scheme is a rational decision plan for choosing both the formulation components and the final dosage form. This decision tree must be flexible enough to accommodate a wide variety of proteins, it must allow for both liquid and solid dosage forms, and it must provide a rationale at each decision point. This framework allows the pharmaceutical scientist to decide what types of formulations are worth considering, both in terms of stability and the demands and desires of the marketing and clinical groups.

Often the design of a formulation of a protein pharmaceutical is not well developed. Companies attempt to expedite formulation selection by neglecting critical issues in formulation and stability. It is not uncommon to find sub-optimal formulations such as phosphate-buffered saline, possibly with the addition of human serum albumin as a ubiquitous protein stabilizer. For a few proteins, this approach has been sufficient, even resulting in commercially successful products, such as erythropoietin and granulocyte colony stimulating factor. However, there is a distinct danger in assuming that such successes with minimal formulations are generally applicable. Moreover, the current regulatory environment, and concerns in the general public surrounding the use of human-derived products in pharmaceuticals, makes the use of human serum albumin increasingly more difficult or even not acceptable. Therefore, there is pressure for any company selling a product that contains albumin to develop an albumin-free formulation. The same pressures are forcing companies to move from animal-derived surfactants to those obtained from vegetable sources. In addition, there is a realization that albumin, being a protein, is not indefinitely stable, and should have an expiration date above and beyond that of the active ingredient. All of these issues affect the list of possible excipients that can be employed in a protein pharmaceutical product.

Constraints on a Pharmaceutically Acceptable Protein Formulation

Before we discuss the choice of dosage forms and excipients in the context of maximizing protein stability, we need to recognize the myriad of constraints placed upon the formulation by the rest of the company and by regulatory bodies. The first constraint, given the global nature of the pharmaceutical industry, is to have excipients that are acceptable in the major markets. This includes avoiding animal-derived additives, as discussed above. It also means that the list of possible additives is effectively limited to those already found in approved products in North America, Europe, and Japan. While this is not an immutable rule, few companies are willing to bear the added cost of getting a new excipient on the market while seeking approval for a new drug product. A major impact of this constraint is to move away from formulations containing albumin, which is still found in a number of protein products. One can anticipate that these older formulations will disappear in the coming years and be reformulated into second-generation products. A recent example has been the development of an albumin-free formulations for recombinant Factor VIII.

The second consideration is the tonicity of the final formulation. If a protein drug is to be administered by intravenous bolus injection or subcutaneously, rather than by continuous infusion, there are strict isotonicity and pH considerations that have to be met for a pain-free injection. Similarly, there are constraints on the choice of excipients, as some (e.g., citrate) have been shown to be irritating upon injection. This is one example of restrictions being set by the clinical groups, which should be clearly defined prior to the start of the formulation process.

Likewise, the eventual route of administration must be taken into account. A convenient administration system is highly desirable for ease of use in the clinical setting, so it is both a marketing and clinical issue. For example, in the case of a subcutaneous injection, there is a maximal volume (~1 ml.) that can be given to a patient without discomfort. Given the expected dose range, this will define the protein concentration that must be used. For other routes of administration, the ability to manipulate the protein concentration as a variable may be possible. Otherwise, it will be fixed, and the decision point regarding protein concentration is removed from the process described below.

The fourth issue is to work with a well-characterized bulk drug substance. This criterion will require cooperation and collaboration with the process science groups involved in fermentation and purification. Variation in the quality of the starting material will compromise the validity of any preformulation or stability work done to date, as well as potentially invalidate data from toxicological tests.

The fifth constraint to be determined is the minimal stability required to complete proof-of-principle investigations, meaning animal studies and initial clinical trials. These investigations can extend for up to one year. Hence, at least a one-year shelf life for even the initial formulation may be required.

Finally, the manufacturing of the protein product must be possible at reasonable costs and allow adaptation to standard manufacturing processes. It is imperative that manufacturing economic analysis be conducted as soon as possible in the drug development process, including an assessment of whether sufficient manufacturing capability exists if the product is successful. Novel formulations that require special manufacturing setups and processes may not offer the advantages of speed which is necessary when one is confronted with development of numerous protein drug candidates. Most likely, specialized dosage forms and/or drug delivery systems will only be investigated with partners who possess the technology and capacity to solve these critical manufacturing issues. In addition, as the in-house pilot plants are challenged with manufacturing multiple products, with limited facilities and resources, there undoubtedly will be pressures to outsource the first clinical manufacturing because of both cost-effectiveness and timescale issues.

Proper Choice of Dosage Form

Considering all the therapeutic protein products on the market, it appears that three dosage forms are viable. The most preferred would be a solution formulation that is typically stored in the refrigerator and preferably in a pre-filled syringe. Second, a frozen solution loaded in a syringe or vial and thawed at the site of administration. Third, a lyophilized formulation, which can be stored at room temperature and reconstituted when needed. Each one of these dosage forms offers various advantages and disadvantages, in the speed of development, manufacturing, packaging and shipment logistics, and in administration of the product to the patient.

Generally, a frozen protein formulation offers the advantages of rapid development provided the protein is stable after multiple freeze/thaw cycles. It is also important to note that freezing can lead to selective precipitation of certain buffer species, producing possibly large pH shifts (van den Berg, 1959; van den Berg and Rose, 1959). This variation in pH, and the potential for damage at the water-ice interface (Chang et al., 1996; Hsu et al., 1995; Krielgaard et al., 1998), might limit the use of such a dosage form. Furthermore, there are disadvantages associated with a frozen formulation concerning the packaging complexities involved in labeling, storage and maintaining the cold chain during transportation of a frozen product. These factors can add significantly to the development costs.

The second option for dosage form is a stable liquid formulation that can be stored in the refrigerator or at room temperature. This dosage form offers the advantages of potentially rapid development and ease of administration. However, for most proteins maintaining physical and chemical stabilities in aqueous solution for an extended period of time is extremely difficult. In addition, because the protein is free in solution there is potential for adsorption to the vial, whether it is made of glass or some type of plastic. A liquid product will have to be shown to be impervious to damage at the air-water interface during transportation, because agitation-induced denaturation and aggregation is a common problem with therapeutic proteins. There is also the possibility of leaching of metals and organics from the container and stopper, which is why some of the vials will need to be inverted during stability testing. These problems have been reported for drug products of all types (Airaudo et al., 1990; Nedich, 1983). Leached materials are particularly problematic with plastic vials and non-Teflon-coated stoppers. With glass vials the leachables could include ionic species that alter the pH and ionic strength of the solution during storage. Finally, the integrity of the container closure must be demonstrated, in order to ensure sterility of the product. For multi-use vials, preservatives can be added, but some are known to foster degradation of proteins (Lam et al., 1997; Rodrigues-Silva et al., 1999). Therefore, development of a stable liquid protein formulation has to be considered very carefully. Determination of the primary pathways responsible for decomposition of the active therapeutic protein is critical for developing this particular dosage form. All of these issues can add significant costs in terms of resources and analytical testing.

Most protein pharmaceuticals currently on the market are sold as lyophilized formulations. Placing proteins in the dried state significantly increases the shelf life, if the proper formulation and lyophilization cycle are used. The combined stresses of drying and freezing are inherently damaging to proteins, and only proper formulation will ensure that a stable product will be obtained. Fortunately, enough is understood now about freezing and drying of proteins that rational design of lyophilized formulations can be undertaken. Moreover, there are some formulations that seem to work well for a variety of proteins. We suggest that part of the strategy for rapidly identifying viable formulations is to make liberal use of some standard protein formulations. The rationale for these compositions has been described in detail in a number of review articles and book chapters, and is discussed in Chapter 5 (MacKenzie, 1976; Franks, 1990; Pikal, 1985, 1994; Carpenter and Chang, 1996; Carpenter et al., 1997). An approach towards developing a lyophilized formulation is described below.

As other formulation possibilities become more developed, one can modify the decision tree to accommodate them. For example, one can imagine alternatives to lyophilization for obtaining solid dosage forms. Both spray-drying and spray-freeze drying are being studied. However, it is expected that the formulation strategy for these processes will be quite similar to that used for lyophilization.

The most likely future additions to the dosage form list will be suspensions and specialized drug delivery systems. Protein suspensions can be produced by a number of different methods. The protein could be dried and suspended in a non-aqueous medium. Alternatively, the protein could be salted-out, using preferentially excluded solutes, such as ammonium sulfate or sucrose, leading to an aqueous vehicle containing most of the protein in its native state but in a solid form, because the solubility limit has been exceeded. The same effect might be accomplished by adjusting the pH to be near the isoelectric point of the protein. For the current version of HTF, we have chosen not to include these dosage forms until their use becomes more widespread and manufacturing facilities are developed to accommodate production of clinical batches.

The same can be said of specialized drug delivery systems, such as polymeric microspheres. Poly lactide/poly glycolide microspheres for sustained delivery of peptides have been in use for years. Recently, reports of their use for proteins, such as human growth hormone, have appeared. These systems bring their own specific stability issues (Cleland, 1998) in addition to making analysis of the stability of the active ingredient more challenging (Yang et al., 1999).

As discussed above, input from various departments, such as discovery research, pharmacokinetics, toxicology, purification, manufacturing, clinical and marketing, must be considered when deciding on the most appropriate dosage form for a given product. In the past, the drug development process has been more or less sequential, going from research to development to clinical, to manufacturing, and then to marketing. A change in strategy is required where there is more interactive, simultaneous communication between the various drug development departments. With the proper information from the other units, one should rapidly be able to decide on dosage form and develop candidate formulations.

Preformulation Studies

The exquisite sensitivity of protein structure, function, and stability to the primary sequence does not readily lend itself to a generic approach for protein formulation. Therefore, determination of the relative importance of various degradation pathways and elucidation of instability mechanisms for a given protein are essential. Even for closely related proteins, the relative stability and major pathways for degradation might be quite different. Thus, the preformulation studies of new protein entities become very important and crucial tasks in the overall HTF strategy.

Preformulation refers to experimental studies designed at understanding the chemical and physical behavior of the protein or drug candidate. The major focus is on the solubility and stability as a function of a number of extrinsic factors, such as pH, protein concentration, ionic strength, buffer composition and temperature. This information will guide the appropriate choice of excipients, buffers and final protein concentration. Although a number of studies could be envisioned, HTF requires that these studies be very specific and focused. Any information that may narrow the choices would be essential for speedy formulation development. Relevant data can be gathered from the process-engineering group responsible for purification or literature studies on similar proteins, even if it is anecdotal. The final dosage form must also include constraints from marketing, clinical and regulatory groups addressing issues related to eventual approval of a pharmaceutically acceptable formulation for market. Ultimately, all of these data may be required to ascertain the most beneficial composition of excipients and the most appropriate dosage form for the protein product in hand.

Typical time scales for preformulation can range from one to three months depending on what assay systems have been established by the research groups and whether they can be utilized as stability-indicating. Therefore, early involvement of all of the formulation experts is critical for rapid development of lead formulations during the developmental phase. An additional driving force is also the fixed time required to generate the stability data to validate the formulation choice.

Proper Choice of Excipients

Liquid Formulations. If a solution dosage form is indicated, then there will be a finite set of possible excipients, restricting choices to those that are found in approved products and have been shown to be effective in protein formulations. For solution formulations, a list of possible excipients is given in Table 2.

Given that most protein formulations will exist at pH values between 4 and 9, there is a limited set of buffers that will exhibit sufficient buffering capacity. A number have been used in protein formulations and they are listed in Table 2. The main criteria for selection are good buffering capacity at the desired pH and lack of ability to accelerate specific chemical and physical reactions. Some proteins can preferentially bind certain buffer species, altering the conformational stability of the protein. In addition, some buffers (e.g., phosphate, Tris) have been

Table 2.
Possible Excipients for Use in Liquid Formulations

Excipient Class	Choices
Buffers	Histidine, Succinate, Acetate, Citrate Phosphate, Tris, Carbonate
Salts	Sodium Chloride, Calcium Chloride Magnesium Chloride
Non-Specific Stabilizers	Sucrose, Trehalose, other sugars, Amino acids (e.g., lysine, glycine)
Specific Stabilizers	depends upon the protein
Surfactants	Tween 20, Tween 80, Pluronic F-68 Sodium Dodecyl Sulfate
Chelators	EDTA

reported to catalyze certain hydrolytic reactions, such as deamidation (Patel et al., 1990).

Because proteins are polyelectrolytes, the ionic strength of the medium can often have an effect on structural stability. The most common salt used in protein formulations is sodium chloride. For an isotonic product, often it is advantageous to keep the concentration of salt as low as possible. Otherwise, the ability to employ stabilizing excipients will be limited.

Structural stabilization of proteins by specific ligands or cofactors in solution formulations can be important for formulation optimization. The mechanism by which these compounds increase thermodynamic stability is described by Timasheff (1992, 1995, 1998) and in other Chapters in this book. Several pharmaceutically important proteins have been stabilized by addition of specific ligands (including metals ions such as calcium), including fibroblast growth factor (Volkin et al., 1993) and DNAse (Chen et al., 1999).

To stabilize proteins (both in aqueous solution and during freezing) with non-specific compounds (e.g., sugars), relatively high concentrations (ca. $> 0.2\,M$) of ligand (solute) are needed to affect protein stability. The mechanism of such stabilization is described in more detail in Chapters 2 and 3. The main point for the current discussion is that in the presence of nonspecific stabilizing excipients fewer protein molecules are unfolded at any point in time, either partially or completely, which reduces the amount of aggregation-competent species and the rate of aggregation (Kendrick et al., 1997, 1998). Addition of small amounts of a specific ligand should theoretically accomplish the same outcome. The most effective non-specific stabilizers tend to be disaccharides, such as sucrose and trehalose. However, certain salts, amino acids, and polymers are preferentially excluded as well. Still, unless there is evidence for advantage in use of a particular compound from this group, sucrose and trehalose should remain the first-line choices.

Finally, one must decide on whether to include a surfactant in the formulation. Surfactants have been shown to inhibit protein aggregation during agitation (e.g., Bam et al., 1995; Krielgaard et al., 1998). The mechanisms by which surfactants stabilize proteins are discussed in a separate Chapter. The surfactant often will be most effective at or above the critical micelle concentration. In Table 2, only the surfactants that have been approved for use in parenteral products in the U.S. have been listed. Note that one potential drawback of including surfactants in the formulation are contaminants (e.g., peroxides) that could cause chemical degradation of proteins (e.g., oxidation).

Lyophilized Formulations. It can be assumed that most proteins will not exhibit sufficient stability in aqueous solution to allow a liquid formulation to be developed. Our understanding of the basic requirements for obtaining a stable lyophilized protein formulation is relatively well developed. The question then is what combination of excipients will allow such a formulation to be prepared. The minimal composition includes a buffer species, an additive capable of forming an amorphous glassy state and inhibiting lyophilization-induced unfolding in which the protein remains entrapped, a bulking agent to provide cake stability, and possibly a surfactant to retard surface-induced damage and/or promote refolding. The rationales for the minimal criteria for a stable lyophilized formation and the choice of appropriate excipients are described in detail in another Chapter.

While a comprehensive formulation development algorithm is yet to be published, following the HTF process outlined above, one can imagine obtaining a rational approach towards selecting final candidate formulation that can be used routinely, even with new classes of therapeutic proteins. Clearly, the details of each case will be guided by the outcomes of the preformulation studies. However, even without a complete data set from the preformulation experiments, implementation of a few standard formulations should be possible. Ultimately, we believe there will be a relatively small number (six of less) of generic, standard lyophilized formulations that will work for nearly all proteins. Each of these will have a corresponding lyophilization cycle that can be used to achieve optimal results. When the HTF process has evolved to this optimal state, the personnel demands be greatly reduced (in terms of full-time equivalents, or FTEs) and the time lines should shrink significantly from the current average of nearly one year.

Estimates of Resources Needed for Formulation Development

The current estimates of the resources required and the possible timelines for formulation development are given in Table 3. Specialized formulations

Table 3.
Estimated Resources Needed for Development of Final Stable Formulations of Protein Pharmaceuticals

Resource	Current Status	Initial HTF Targets	Optimal HTF Targets
• Personnel			
Standard Formulations	2 FTE	2 FTE	2 FTE
Complex Formulations	3–5 FTE 3–4 FTE	3 FTE	
• Time Scale			
Standard Formulations	12 months	6–9 months	3–6 months
Complex Formulations	12–24 months	12 months	6–9 months
• Minimal Amount of Material			
Standard Formulations	200–500 mg	100–200 mg	50–100 mg
Complex Formulations	400–2000 mg	200–500 mg	100–200 mg

(suspensions and controlled release dosage forms) will require more resources and less aggressive time lines (see Table 3). However, even this situation should improve as more of these types of products enter the marketplace. We anticipate that a significant number of these types of products will be approved over the next ten years. By then, most large pharmaceutical companies will have acquired the necessary expertise to develop them at rates comparable to the standard formulations of today. Once the HTF scheme has evolved to this stage, a company should be able to develop final formulations quickly and easily, even if personnel levels are relatively constant.

Interestingly, while a streamlined approach towards final formulation selection is the heart of the HTF strategy, it is the one aspect that can be introduced into any current product development scheme almost immediately, as it can be accomplished solely based on the current state of understanding of protein stabilization. Other aspects of the HTF package will demand more time to implement, as they will require significant advances to be made, particularly in the areas of analytical methodology and information technology. These two features of the HTF scheme are described below.

USE OF SOFTWARE AND DATABASES TO ASSIST IN THE HTF PROCESS

Unless formulation scientists have access to all pertinent information, critical choices for formulation components could be delayed or, even worse, incorrect. Therefore, we propose that a key aspect of the HTF process is to develop an integrated information system that can assist the scientist in making critical decisions. Each and every step in the formulation development process is predicated upon having the most recent and comprehensive information on the physicochemical properties of the protein under development, as well as on the effects of additives on protein structure and stability. Clearly, the latter part of this goal is ever evolving as our knowledge of protein chemistry increases. Therefore, the final HTF scheme will involve a central database of essential information. While a large amount of this type of information is currently available, it appears in a multitude of sources, generated by researchers in disparate disciplines, ranging from materials science to biochemistry to food science to pharmaceutics. This makes access to the data difficult. Unless efforts are made to assemble this data in a cohesive package soon, the challenge will become even greater. Once the task is accomplished, sufficient information should be available to predict the performance of certain types of formulations.

In the absence of such resources, one can still make use of existing software and databases. These are summarized below. Briefly, they cover three areas: predictive protein algorithms based upon primary sequence, sequence homology analysis software, and programs aimed at predicting relative stability of proteins. Summaries of the locations and availability of these programs and databases has recently been published (Cook, 1999; Apweiler, 1999).

Predictive algorithms allow one to analyze a primary sequence and predict a variety of chemical and physical properties. For example, there has been enormous effort to develop schemes whereby one can predict the three-dimensional folding of a protein based solely on its sequence. Although there have been some successes, it has been demonstrated in double-blind studies that analysis of new sequences, when they have no clearly related partner in the existing database, fail to provide an accurate prediction of the folded conformation (Orengo et al., 1999; Vencloyas et al., 1999). When the analysis can be guided by sequence homology, the accuracy rises dramatically. Likewise, one can imagine developing structure-stability relationships, based solely upon primary sequence. Some studies of this nature have been reported. For example, Pandit and co-workers relate the frequency of certain dipeptide sequences to instability in proteins (Reddy, 1996; Guruprasad et al., 1990). The result is an algorithm that predicts the relative stability of a protein. Although the degradation criteria are based upon intracellular stability, this approach does provide researchers with an expectation of the overall fragility of the protein in question.

In another example of this type of sequence analysis, Roger et al. (1986) identified regions in proteins that were high in proline, acidic side chains (Asp and Glu), serine, and threonine. The presence of these regions was associated with more rapid degradation *in vivo*. Similar correlations should exist for *in vitro* stability. Even programs to predict likely deamidation sites (where Asn is converted into Asp-like residues) would be helpful, especially as our ability to remove those sites by recombinant DNA technology improves. There is now a solid basis on which to predict both primary and secondary structure effects on deamidation rates (Wright, 1991; Xie and Schowen, 1999), which is considered in more detail in Chapter 4.

Given that numerous protein sequences, whether determined directly or extrapolated from nucleotide data, are available in databases that can be accessed via the world wide web, focused studies generating data on specific instabilities (e.g., deamidation, oxidation, aggregation) could lead to similar structure-stability relationships. The number of available sequences has been doubling every 18–24 months over the last ten years (Cook, 1999). Similarly, there has been an explosion in algorithms and software to analyze these data. Many focus on sequence homology and identification of functional and genetic relationships. Given these advances, it is not unreasonable to expect that stability-indicating profiles could be developed as well.

In addition to stability predictions, many programs are available to calculate important chemical and physical properties such as the isoelectric point, the hydrophobicity profile, the positioning of secondary structure elements, and the conformational flexibility. Knowledge of these properties would help the formulation scientist anticipate the behavior of a new molecule provided for development into a potential drug product. Furthermore, algorithms to predict correlations of these physical properties with physical and chemical stabilities of therapeutic proteins should be straightforward to design and implement with current programs.

ESSENTIAL ANALYTICAL METHODS

One must have access to certain equipment in order to characterize properly protein degradation pathways and effects of formulation variables on protein stability. In addition, we will discuss some areas where technological advancements must be made in order to allow formulation scientists to work with smaller amounts of protein. While a number of reliable and accurate stability-indicating assays exist, there needs to be efforts to miniaturize some of them. This will allow development of proteins where the amount of material is limited, e.g., less than 100 milligrams total.

The first analytical methods that are needed are those that can quantify the protein concentration. As determination of accurate extinction coefficients is straightforward, ultraviolet absorption spectroscopy is inexpensive and preferred over chemical methods, such as the Lowry method. One also needs to be able to monitor the global structure of the protein. Circular dichroism (CD) spectroscopy allows one to follow changes in both secondary and tertiary structure of proteins in solution (Manning, 1994). Infrared spectroscopy can also be used to detect secondary structural changes (see below), and fluorescence and derivative UV spectroscopies can be used to study tertiary structure.

The second category of required analytical methods are those that monitor retention of native primary structure, namely, chromatographic methods, such as reversed phase HPLC and ion exchange HPLC, or alternative methods like capillary electrophoresis. Among the most useful applications of these HPLC methods is the generation of a "tryptic map", which is accomplished by digesting the protein with a sequence-specific enzyme (e.g., trypsin) and separating the resulting peptide fragments by reversed phase HPLC. The pattern then becomes a sensitive fingerprint of the protein. Any chemical modification of the protein will cause one or more of the peaks to shift, indicating damage to the protein. Mass spectrometry can be used to determine the specific chemical modification leading to the peak shift.

Among the most important degradation pathways for proteins is aggregation. A number of methods are available to characterize and quantify aggregation levels in proteins, including polyacrylamide gel electrophoresis (PAGE), size exclusion chromatography (SEC), light scattering techniques and analytical ultracentrifugation. Although each of these methods has its advantages and disadvantages, all four should be considered essential, as they can be complementary. They allow both qualitative and quantitative characterization of the soluble and insoluble aggregates in a bulk substance or final formulation.

For lyophilized formulations, a certain set of equipment is considered to be the minimum required for conducting formulation development properly. These include an infrared (IR) spectrometer, a Karl Fischer titrator, a differential scanning calorimeter (DSC), and x-ray diffractometer. The IR instrument will allow one to determine the extent of structural damage (at a secondary structure level) in the solid state. In addition, it can be used to monitor the secondary structure composition in the original solution (prior to lyophilization), in the frozen solid and in the reconstituted product (Carpenter et al., 1998). X-ray diffraction of solids allows one to determine the crystallinity of a solid sample. In formulations where the bulking agent is crystallizing during the lyophilization process, one needs to know whether the crystallization is complete. Incomplete crystallization can lead to degradation during storage (cf. Carpenter et al., 1997). The Karl Fischer apparatus allows one to measure the amount of water in the final cake. Typically, moisture contents need to be at 1–2 % in order for the product to have the optimal glass transition temperature (T_g) and maximal stability. Finally, DSC is required to measure the T_g of the final formulation. The storage temperature must below the T_g. As the minimal acceptable storage temperature for lyophilized formulations is room temperature, a minimum T_g would be about 40° C. In addition to these instruments, access to a microscope equipped with a lyophilization stage is valuable.

The issue of how to develop formulations with limited amounts of material is a critical concern. One can imagine adapting a number of existing technologies to assist in assessing the viability of a given formulation while using very small amounts of material. These would include:

1. microtiter plate assays,
2. robotics to assist in sample preparation and analysis,
3. ELISA or other antibody-based recognition assays for intact drug,
4. multi-sample adaptations of existing analytical techniques (e.g., DSC or analytical ultracentrifuge)
5. nanoseparation methodology (the so-called "lab-on-a-chip" approach),
6. surface-immobilized analytical methods, such as the BIAcore.

Each of these methods could provide significant increases in the capability to monitor the stability of a protein in a microformulation (less than 10 µg of active drug). For example, new chip technologies (either DNA- or protein-based) are becoming powerful tools that are increasingly used in all areas of drug discovery. They offer significant advantage over older technologies in terms of speed, sensitivity and cost. It is important to consider the use or implementation of these tools in formulation studies. One can envision protein chips doing for the study of proteins what DNA chips have done to the study of genes. That is, they can be used to identify critical parameters required for rapidly developing a stable formulation. Similarly, any of the other methods could be modified to assist the formulation scientist.

Finally, there has been a steady improvement in protein analysis tools such as mass spectrometry and more powerful and versatile biochemical techniques, which are not only more sensitive but offer the advantages of speed and low cost. So with the emerging technological advances in analytics and implementation of these enabling technologies in preformulation and formulation studies, one can envision a more time and resource efficient process for developing stable protein pharmaceuticals. Most of these technologies are being developed to meet the needs of efficient drug discovery research. The challenge here will be the speed at which the formulation scientists adapt these technologies into their programs. Unfortunately, these tools have not yet reached their maturity, so the implementation of micro-analytical methods for preformulation and formulation is crucial for developing HTF strategies for protein pharmaceuticals.

STABILITY PROTOCOLS

In addition to formal storage stability protocols that would be acceptable to regulatory agencies for real-time data, one must have standard approaches for assessing the possible long-term fate of the candidate formulations. Given the aggressive time line of the HTF strategy, one must rely, to some degree, upon accelerated storage studies to guide decision about suitable excipients. These could be done as isothermal studies, storing the protein at room temperature and 40° C, for example. In certain cases, exposure to higher temperatures, such as 50° C or even 60° C, may be warranted. However, it is important to note that at higher temperatures, there is the distinct possibility that the rate-determining pathway may change compared to that at the projected storage conditions.

An alternative to isothermal testing that needs to be considered is the use of non-iosothermal stress testing. In this procedure, the sample is exposed to a linear increase in temperature over time (or approximately linear with respect to time). Samples are taken at set intervals and assayed for protein damage. Using an algorithm developed by Lee and Stavchansky (1998), it is possible to determine the Arrhenius parameters in a single experiment, which would otherwise require at least three separate isothermal studies. This approach has been demonstrated for a pentapeptide (Lee and Stavchansky, 1998), and should be extended to larger proteins in the future.

UNIFIED STRATEGY FOR HTF

The requirements for HTF can be summarized in four points. First, integration of speed, sensitivity and cost can be achieved by having in place good analytical methodologies for monitoring both chemical and physical instability mechanisms. These methods can serve as tools for elucidating the mechanism of instability of protein drugs and also as stability indicating assays during analysis of the final formulation. Second, identification of the instability mechanism of proteins can lend itself to developing a fast-track stable liquid protein formulation, although the disadvantages of addressing biocompatibility and transportation issues have to be addressed with such dosage forms. A third requirement for HTF is to have obviously a safe and a convenient dosage form for administration. Coupled with this criterion should be the requirement for manufacturing the formulation using standard processes, so that if needed the production can be out-sourced to a contract facility.

Finally, technological advances will be a major driver for HTF and in general biotechnology development. The ability to work at low microgram to milligram scale of protein is essential for the ability to start the process before the drug candidate achieves development status. In many cases, companies may initially neglect formulation and stability issues that will haunt them in later stages of development. The development of successful formulations is dependent upon the ability to study both the *in vitro* and *in vivo* characteristics of the protein drug, as well as its intended application in the clinic. Thus, it is essential for the formulation group to work closely with the discovery research, the pharmacokinetics and the toxicology departments prior to the decision to move the drug into full-scale development. Resolving the difficulties in the early research and development stages will not only help in targeting the appropriate drug candidate for development, but also speed up the selection of a pharmaceutically acceptable formulation for clinical trials. What we propose is the involvement of the formulation group in a multi-disciplinary environment where interactions start early in the research environment. The appropriate lead candidates would be selected not only based on efficacy studies, but also on pre-formulation studies. We envision these studies to take one to three months to establish appropriate profiles for formulation development. As a result, the lead formulation can have the advantages of demonstrating efficacy and potential stability of the protein drug.

During the development process of a biotechnology drug we envision formulation to take typically three to six months. This period would encompass preformulation studies and evaluation of the formulation options determined by the planned clinical studies. Hence, protein formulation and purification can be developed concurrently using limited quantities of the product, and identification of the lead formulation could occur at the time of establishment of the purification process. What we propose is not a sequential development pathway for protein drugs, but a concurrent scenario where formulation design is initiated very early in the drug development pipeline. We are quite confident that the HTF can be coupled successfully with high throughput screening strategies of protein drugs and facilitate drug development within the inherent time and resource constraints of the pharmaceutical industry.

REFERENCES

Airaudo, C.B., Gayte-Sorbier, A., Momburg, R., and Laurent, P., 1990. Leaching of antioxidants and vulcanization accelerators from rubber closures into drug preparations, *J. Biomater. Sci., Polymer Ed.* 1:231.

Aldous, B.J., Auffret, A.D., and Franks, F., 1995. The crystallization of hydrates from amorphous carbohydrates, *Cryo-Letters* 16:181.

Anchordoquy, T.J. and Carpenter, J.F., 1996. Polymers protect lactate dehydrogenase during freeze-drying by inhibiting dissociation in the frozen state, *Arch. Biochem. Biophys.* 332:231.

Apweiler, R., 1999. Introduction to molecular biology databases, in: *EBI Online Manual on Molecular Biology Databases*, R. Apweiler, R. Lopez, and B. Marx, eds., URL: www.ebi.ac.uk/swissprot/Publications/mbd1.html.

Bam, N., Cleland, J., and Randolph, T.W., 1995. Stability of protein formulations: investigation of surfactant effects by a novel EPR spectroscopic technique, *Pharm. Res.* 12:1.

Bam, N., Cleland, J.L., Yang, J., Carpenter, J.F., Manning, M.C., and Randolph, T.W., 1998. Tween protects recombinant human growth hormone against agitation-induced damage via hydrophobic interactions, *J. Pharm. Sci.* 87:1554.

Bell, L.N., Touma, D.E., White, K.L., and Chen, Y.H., 1998. Glycine loss and Maillard browning as related to the glass transition in a model food system, *J. Food Sci.* 63:625.

Carpenter, J.F. and Chang, B.S., 1996. Lyophilization of protein pharmaceuticals, in *Biotechnology and Biopharmaceutical Manufacturing, Processing, and Preservation*, K.E. Avis and V.L. Wu, eds., Interpharm Press, Buffalo Grove, IL.

Carpenter, J.F., Pikal, M.J., Chang, B.S., and Randolph, T.W., 1997. Rational design of stable lyophilized protein formulations: some practical advice, *Pharm. Res.* 14:969.

Carpenter, J.F., Prestrelski, S.J., and Dong, A.,1998. Application of infrared spectroscopy to development of stable lyophilized protein formulations, *Eur. J. Pharm. Biopharm.* 45:231.

Chang, B.S. and Randall, C.S., 1992. Use of subambient thermal analysis to optimize protein lyophilization, *Cryobiology* 29:632.

Chang, B.S., Kendrick, B.S., and Carpenter, J.F., 1996. Surface-induced denaturation of proteins during freezing and its inhibition by surfactants, *J. Pharm. Sci.* 85:1325.

Chen, B., Costantino, H.R., Hsu, C.-C., and Shire, S.J., 1999. Influence of calcium ions on the structure and stability of recombinant human deoxyribonuclease I in the aqueous and lyophilized States, *J. Pharm. Sci.* 88:477.

Cleland, J.L., 1998. Solvent evaporation processes for the production of controlled release biodegradable microsphere formulations for therapeutics and vaccines, *Biotechnology Progress* 14:102.

Cleland, J.L., Powell, M.F., and Shire, S.J., 1993. The development of stable protein formulations: a close look at protein aggregation, deamidation, and oxidation, *Crit. Rev. Ther. Drug Carrier Systems* 10: 307.

Cook, J.L., 1999. Internet biomolecular resources, *Anal. Biochem.* 268:165.

Crowe, J.H., Carpenter, J.F., and Crowe, L.M., 1998. The role of vitrification in anhydrobiosis. *Ann. Rev. Physiol.* 60:73.

Franks, F., 1990. Freeze drying: from empiricism to predictability, *Cryo-Letters* 11:93.

Guruprasad, K., Reddy, B.V.B., and Pandit, M.W., 1990. Correlation between stability of a protein and its dipeptide composition: a novel approach for predicting *in vivo* stability of a protein from its primary sequence, *Protein Eng.* 4:155.

Heller, M.C., Carpenter, J.F., and Randolph, T.W., 1999. Application of a thermodynamic model to the prediction of phase separations in freeze-concentrated formulations for protein lyophilization, *Arch. Biochem. Biophys.* 363:191.

Hsu, C.C., Nguyen, H.M., Yeung, D.A., Brooks, D.A., Koe, G.S., Bewley, T.A., and Pearlman, R., 1995. Surface denaturation at solid-void interface- a possible pathway by which opalescent particulotes form during the storage of lyophilized tissue-type plasminogen activator at high temperatures, *Pharm. Res.* 12:69.

Kendrick, B.S., Chang, B.S., Arakawa, T., Peterson, B., Randolph, T.W., Manning, M.C., and Carpenter, J.F., 1997. Preferential exclusion of sucrose from recombinant interleukin-1 receptor antagonist: role in restricted conformational mobility and compaction of native state, *Proc. Natl. Acad. Sci. USA* 94:11917.

Kendrick, B.S., Carpenter, J.F., Cleland, J.L., and Randolph TW., 1998. A transient expansion of the native state precedes aggregation of recombinant human interferon-gamma, *Proc. Natl. Acad. Sci. USA* 95:14142.

Koseki, T., Kitabatake, N., and Doi, E., 1990. Freezing denaturation of ovalbumin at acid pH, *J. Biochem.* 107:389.

Krielgaard, L., Jones, L. S., Randolph, T.W., Frokjaer, S., Flink, J.M., Manning, M.C., and Carpenter, J.F., 1998. Effect of Tween 20 on freeze-thawing and agitation-induced aggregation of recombinant human factor XIII, *J. Pharm. Sci.* 87:1597.

Lam, X.M., Patapoff, T.W., and Nguyen, T.H., 1997. The effect of benzyl alcohol on recombinant human interferon-gamma, *Pharm. Res.* 14:725.

Lee, J.C. and Timasheff, S.N., 1981. The stabilization of proteins by sucrose, *J. Biol. Chem.* 259:7193.

Lee, M.-L. and Stavchansky, S., 1998. Isothermal and nonisothermal decompositon of thymopentin and its analogs in aqueous solution, *Pharm. Res.* 15:1702.

Li, S., Schoneich, C., and Borchardt, R.T., 1995. Chemical instability of protein pharmaceuticals: mechanisms of oxidation and strategies for stabilization. *Biotechnol. Bioeng.* 48:490.

MacKenzie, A.P., 1976. The physico-chemical basis of the freeze-drying process, *Dev. Biol. Stand.* 36:51.

Manning, M.C., Patel, K., Borchardt, R.T., 1989. Stability of protein pharmaceuticals, *Pharm. Res.* 6:903.

Nedich, R.L., 1983. Selection of containers and closure systems for injectable products, *Am. J. Hosp. Pharm.* 40:1924.

Orengo, C.A., Bray, J.E., Hubbard, T., LoConte, L., and Sillitoe, I., 1999. Analysis and assessment of ab initio three-dimensional prediction, secondary structure, and contacts prediction, *Proteins: Struct., Func., Gen. Suppl.* 3:149.

Patel, K. and Borchardt, R.T., 1990. Chemical pathways of peptide degradation. II. kinetics of deamidation of an asparaginyl residue in a model hexapeptide, *Pharm. Res.* 7:703.

Pearlman, R. and Nguyen, T.H., 1992. Pharmaceutics of protein drugs, *J. Pharm. Pharmacol.* 44(Suppl 1):178.

Pharmaceutical Research and Manufacturers of America, 1999. PhRMA Annual Survey. URL: www.phrma.org.

Pikal, M.J., 1985. Use of laboratory data in freeze drying process design: Heat and mass transfer coefficients and the computer simulation of freeze-drying, *J. Parent. Drug Assoc.* 39:115.

Pikal, M.J., 1994. Freeze-drying of proteins, in: *Formulation and Delivery of Proteins and Peptides* (J.L. Cleland and R. Langer, Eds.) ACS Symposium Series. 567: 120.

Reddy, B.V., 1996. Structural distribution of dipeptides that are identified to be determinants of intracellular protein stability, *J. Biol. Struct. Dyn.* 14:201.

Rodrigues-Silva, R., Antunes, G.F., Velarde, D.T., and Santoro, M.M., 1999. Thermal stability studies of hyperimmune horse antivenoms, *Toxicon* 37:33.

Tanaka, T. Takeda, T., and Miyajama, R., 1991. Cryoprotective effect of saccharides on denaturation of catalase during freeze-drying, *Chem. Pharm. Bull.* 39:1091.

Timasheff, S.N., 1992. Stabilization of protein structure by solvent additives, in: *Stability of Protein Pharmaceuticals. Part B. In Vivo Pathways of Degradation and Strategies for Protein Stabilization*, T.J. Ahern and M.C. Manning, eds., Plenum Press, New York.

Timasheff, S.N., 1995. Preferential interactions of water and cosolvents with proteins, in: *Protein-Solvent Interactions*, R.B. Gregory, ed., Marcel Dekker, New York.

van den Berg, L., 1959. The effect of addition of sodium and potassium chloride to the reciprocal system: KH_2-PO_4-Na_2HPO_4-H_2O on pH and composition during freezing, *Arch. Biochem. Biophys.* 84: 305.

van den Berg, L. and Rose, D., 1959. The effect of freezing on the pH and composition of sodium and potassium solutions: The reciprocal system KH_2-PO_4-Na_2HPO_4-H_2O, *Arch. Biochem. Biophys.* 81:319.

Vencloyas, C., Ginalski, K., and Fidelis, K., 1999. Addressing the issue of sequence-to-structure alignments in comparative modeling of CASP3 target proteins. *Proteins: Struct., Func., Gen.* Suppl. 3:231.

Volkin, D.B., Verticelli, A.M., Marfia, K.E., Burke, C.J., Mach, H., and C.R. Middaugh, C.R., 1993. Sucralfate and soluble sucrose octasulfate bind and stabilize acidic fibroblast growth factor, *Biochim. Biophys. Acta* 1203:18.

Wright, H.T., 1991 Sequence and structure determinants of the nonenzymatic deamidation of asparagine and glutamine residues in proteins, *Protein Eng.* 4:283.

Xie, M., and Schowen, R.L., 1999. Secondary structure and protein deamidation, *J. Pharm. Sci.* 88:8.

Yang, T.H., Dong, A., Meyer, J., Johnson, O.L., Cleland, J.L., and Carpenter, J.F., 1999. Use of infrared spectroscopy to assess secondary structure of human growth hormone within biodegradable microspheres, *J. Pharm. Sci.* 88:161.

Index

Pharmaceutical Biotechnology

Chronological Listing of Volumes

Volume 1 PROTEIN PHARMACOKINETICS AND METABOLISM
Edited by Bobbe L. Gerraiolo, Marjorie A. Mohler, and Carol A. Gloff

Volume 2 STABILITY OF PROTEIN PHARMACEUTICALS, Part A: Chemical and Physical Pathways of Protein Degradation
Edited by Tim J. Ahern and Mark C. Manning

Volume 3 STABILITY OF PROTEIN PHARMACEUTICALS, Part B: *In Vivo* Pathways of Degradation and Strategies for Protein Stabilization
Edited by Tim J. Ahern and Mark C. Manning

Volume 4 BIOLOGICAL BARRIERS TO PROTEIN DELIVERY
Edited by Kenneth L. Audus and Thomas J. Raub

Volume 5 STABILITY AND CHARACTERIZATION OF PROTEIN AND PEPTIDE DRUGS: Case Histories
Edited by Y. John Wang and Rodney Pearlman

Volume 6 VACCINE DESIGN: The Subunit and Adjuvant Approach
Edited by Michael F. Powell and Mark J. Newman

Volume 7 PHYSICAL METHODS TO CHARACTERIZE PHARMACEUTICAL PROTEINS
Edited by James N. Herron, Win Jiskoot, and Daan J. A. Crommelin

Volume 8 MODELS FOR ASSESSING DRUG ABSORPTION AND METABOLISM
Edited by Ronald T. Borchardt, Philip L. Smith, and Glynn Wilson

Volume 9 FORMULATION, CHARACTERIZATION, AND STABILITY OF PROTEIN DRUGS: Case Histories
Edited by Rodney Pearlman and Y. John Wang

Volume 10 PROTEIN DELIVERY: Physical Systems
Edited by Lynda M. Sanders and R. Wayne Hendren

Volume 11 INTEGRATION OF PHARMACEUTICAL DISCOVERY AND DEVELOPMENT: Case Histories
Edited by Ronald T. Borchardt, Roger M. Freidinger, Tomi K. Sawyer, and Philip L. Smith

Volume 12 MEMBRANE TRANSPORTERS AS DRUG TARGETS
Edited by Gordon L. Amidon and Wolfgang Sadée

Volume 13 RATIONAL DESIGN OF STABLE PROTEIN FORMULATIONS: Theory and Practice
Edited by Mark C. Manning and John F. Carpenter

Volume 14 DEVELOPMENT AND MANUFACTURE OF PROTEIN PHARMACEUTICALS
Edited by Steven L. Nail and Michael J. Akers